Time-Domain Beamforming and Blind Source Separation

Speech Input in the Car Environment

Lecture Notes in Electrical Engineering

Volume 3

Time-Domain Beamforming and Blind Source Separation
Julien Bourgeois, and Wolfgang Minker
ISBN 978-0-387-68835-0, 2007

Digital Noise Monitoring of Defect Origin
Telman Aliev
ISBN 978-0-387-71753-1, 2007

Multi-Carrier Spread Spectrum 2007
Simon Plass, Armin Dammann, Stefan Kaiser, and K. Fazel
ISBN 978-1-4020-6128-8, 2007

Julien Bourgeois • Wolfgang Minker

Time-Domain Beamforming and Blind Source Separation

Speech Input in the Car Environment

 Springer

Julien Bourgeois
Universität Ulm
Institut für Informationstechnik
Albert-Einstein-Allee 43
89081 Ulm
Germany

Wolfgang Minker
Universität Ulm
Institut für Informationstechnik
Albert-Einstein-Allee 43
89081 Ulm
Germany

ISBN 978-0-387-68835-0   e-ISBN 978-0-387-68836-7
DOI 10.1007/978-0-387-68836-7
Springer Dordrecht Heidelberg London New York

Library of Congress Control Number: 2007932057

Printed on acid-free paper

Springer is part of Springer Science+Business Media (www.springer.com)

Preface

The development of computer and telecommunication technologies led to a revolution in the way that people work and communicate with each other. One of the results is that large amount of information will increasingly be held in a form that is natural for users, as speech in natural language. In the presented work, we investigate the speech signal capture problem, which includes the separation of multiple interfering speakers using microphone arrays.

Adaptive beamforming is a classical approach which has been developed since the seventies. However it requires a double-talk detector (DTD) that interrupts the adaptation when the target is active, since otherwise target cancelation occurs. The fact that several speakers may be active simultaneously makes this detection difficult, and if additional background noise occurs, even less reliable. Our proposed approaches address this separation problem using continuous, uninterrupted adaptive algorithms. The advantage seems twofold: Firstly, the algorithm development is much simpler since no detection mechanism needs to be designed and no threshold is to be tuned. Secondly, the performance may be improved due to the adaptation during periods of double-talk.

In the first part of the book, we investigate a modification of the widely used NLMS algorithm, termed Implicit LMS (ILMS), which implicitly includes an adaptation control and does not require any threshold. Experimental evaluations reveal that ILMS mitigates the target signal cancelation substantially with the distributed microphone array. However, in the more difficult case of the compact microphone array, this algorithm does not sufficiently reduce the target signal cancelation. In this case, more sophisticated blind source separation techniques (BSS) seem necessary.

The second part is dedicated to blind separation techniques, much more recent than classical adaptive beamforming (the first results with real acoustic mixings appearing in the nineties). Our objective was to evaluate the performance of blind separation techniques relative to that of more mature beamforming approaches. In addition, we wanted to combine the advantages of beamforming, notably its performance and robustness, with those of blind source separation which does not require activity detection. Parra's frequency-domain block-diagonalization algorithm served as a benchmark in the field of blind source separation. However, we realized that this algorithm could not be flexibly combined with beamformers because it fails to cope with certain "acausal" type of source mixing. We therefore focused on the time-domain approach by Buchner. This approach has been extended to be applied in case there are more microphones than sources. At a moderate computational cost, the proposed Partial BSS scheme flexibly exploits all microphone signals and provides multiple interferer references. Moreover, we derive self-closed update rules that emerge as very robust relative to other algorithms in an experimental comparison. An emphasis is also placed on the theoretical study of BSS, evidencing the role of the causality of the mixing system.

In the last part of the book, we combine both, the beamforming and BSS approaches. While the input of geometrical prior information may increase the start-up performance, we show that the performance gain after the initial convergence is limited. The use of an adaptive interference canceler as a postprocessor leads to a higher interference suppression and also to a higher cancelation of the desired signal. However, we will see that the cancelation of the desired signal may be kept moderate by adequately combining BSS with the ILMS algorithm *and* geometrical prior information.

The presented book results from a cooperation between DaimlerChrysler and the University of Ulm. The industrial partner provided the privileged application field of the car environment and we applied two different, plausible experimental settings using compact and distributed microphone arrays. However, the proposed methods are quite general and should be easily portable to other environments and to different applications.

Acknowledgments

This project would not have been possible without the support of many people.

The authors would like to express their deepest gratitude to Jürgen Freudenberger from the University of Applied Sciences in Konstanz, who read the draft since the early revisions, was always open for discussions and provided numerous advices.

Klaus Linhard from DaimlerChrysler Research and Technology substantially supported the work by his profound technical experience and assistance.

Walter Kellermann from the University of Erlangen took the time to read the manuscript. His valuable comments and constructive criticism helped to make sense of the confusion.

Thanks to Guillaume Lathoud for the many stimulating discussions and his very sound feedback.

Thanks to Jason Ward and Caitlin Womersley from Springer for their assistance during the publishing process.

The authors acknowledge the financial support provided through the HOARSE project from the European Community's Human Potential Programme.

Contents

1

Introduction

Speech is a natural and therefore a privileged communication modality for humans. For example in cars, safety and convenience issues require hands-free (or "seamless") speech-based human–machine interfaces for the driver to manipulate complex functionalities and devices while driving. Applications include hands-free phone calls as well as more advanced functions such as automatic dialog systems for in-vehicle navigation assistance systems [71]. With a seamless speech input, such interfaces increase comfort but have to face several issues:

(i) The signal-to-noise ratio (SNR) at a given microphone can be weak relative to the background noise since the signal energy is inversely proportional to the square of the distance to the sound source [14]. Moreover, room acoustics leads to a reverberated speech signal.

(ii) Interferences, such as speech from the codriver, may greatly hamper the speech recognizer performance, which is crucial for human–machine dialog applications. Separation of the target speaker during periods of competing speech from the codriver represent a particular challenge. This is because the characteristics of the interferer signals cannot be directly estimated from the microphone signals during these periods [50]. This problem is of particular importance since spontaneous multiparty speech contains lots of overlaps between the speech flows of the participants [43].

These issues make the seamless speech input a challenging problem. Before recognizing speech as a sequence of words, an important preprocessing step is to denoise the speech signal from its perturbations. In this book, we address the issue of separating the desired signal from interfering speech, i.e., the point (ii) above.

The car interior, which is weakly reverberant, will be our test environment. Our experiments have been carried out in a Mercedes S320 vehicle with two different microphone arrays: a four-element compact array mounted in the rear-view mirror, and a two-element distributed array mounted on the car

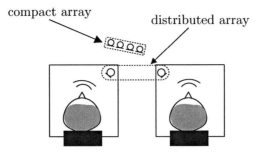

Fig. 1.1. Experimental setup layout for the car interior. We want to separate the driver speech from the codriver interfering speech with a four-element compact array mounted in the rear-view mirror or a two-element distributed array mounted on the car ceiling

ceiling. In this test environment, we want to suppress the codriver interfering speech to recover the clean driver speech, as shown in Fig. 1.1.

1.1 Existing Approaches: A Brief Overview

With a single microphone, noise reduction algorithms rely solely on the temporal (or spectral) information contained in the input signal. They can be effective if the noise spectral content varies slowly relative to the signal but they generally yield a distortion of the target signal. They are not appropriate to suppress local nonstationary interference, such as the codriver speech [66]. For this task, microphone array processing techniques are especially well-suited, since they are able to suppress the local interferer while keeping the desired signal undistorted [17, 50].

Several microphone array methods have been developed in the two last decades. This section provides a brief overview of the methods that can be applied to the separation of the target signal from the interfering speech.

A class of methods, usually referred to as beamforming methods, is based on the prior knowledge of the position of the target speaker. Many fixed (data-independent) beamformers as well as statistically optimum or adaptive beamformers, required for nonstationary noise fields, can be formulated in the linearly constrained minimum variance (LCMV) framework [36, 42, 47, 88].

Adaptive beamformers need to be carefully used in speech enhancement. In practice, inaccurate prior knowledge of the target speaker position or room reverberation can lead to the cancelation of the target speech signal [91]. For this reason, adaptive beamformers should be supervised (or "controlled") so that the adaptation occurs only when the target signal level is weak relative to that of the interferers [27, 54, 50]. In the context of interfering speech signals, the control mechanism is called a double-talk detector (DTD). For most existing adaptive beamformers, the DTD performs in an all-or-nothing manner, that is, the adaptation is interrupted during periods of double-talk.

Even though several approaches have been proposed [44, 49, 54], the design of the DTD appears delicate. It implies tunable parameters, whose value may influence the performance significantly. Also, background noise and nonstationary local interferences create alternating noise fields making the design of a reliable and accurate DTD even more difficult. Therefore, alternative unsupervised techniques that do not require any tunable threshold are of interest, not only because they may be easier to design but also because they are able to adapt during double-talk.

More recently, a class of unsupervised methods has been developed and applied to audio signal processing. These methods, usually referred to as blind source separation (BSS), are based on the sole assumption that the sources are mutually independent. BSS algorithms pursue a similar goal as beamforming: to reduce interferences. A major difference to beamforming is that BSS does not require any information about the target position, and BSS is sometimes termed "blind beamforming" [21]. Another difference is that BSS recovers all present sources simultaneously, while beamforming extracts only one target source.

The premises of BSS can be traced back to the work of Jutten et al. [59], who devised a method to separate instantaneous mixtures. The separation of multiple speakers, an important application of BSS in audio signal processing, is considerably more challenging because it involves convolutive mixtures. Applying BSS to realistic scenarios in audio revealed difficulties [86]. On the one hand, time-domain BSS seems to suffer from very slow convergence [74]. On the other hand, the performance of narrowband frequency-domain BSS is fundamentally limited [9]. Narrowband frequency-domain BSS suffers from the so-called permutation and circularity problems, which require extra repair measures [57, 79, 81, 82, 83]. Even though BSS has been applied successfully in some realistic scenarios [4, 75], its performance in terms of interference suppression is usually regarded as inferior to that of LCMV beamforming, apart from the target signal cancelation problem [11].

Taken individually, the beamforming approach and the BSS approach have revealed their particular drawbacks. In this book, we consider them as complementary and address the question whether they can be combined in an efficient manner.

1.2 Scope and Objective of the Book

The first objective of the book is to evaluate the performance of time-domain LCMV beamforming and BSS algorithms. The evaluation is performed in terms of reduction of the interference signal level first and then in terms of reduction of the word error rate when used as an acoustical front-end to a speech recognizer. The focus is on time-domain signal processing, as opposed to frequency-domain signal processing, for the following reasons:

- Many frequency-domain algorithms rely on an approximative narrowband signal model which ignores coupling across different frequency bins [61]. Not relying on the narrowband signal model may improve the performance and support the understanding of the algorithm.
- Time-domain BSS algorithms seem robust against the permutation problem.

The car environment is a good example of an application where the position of the target speaker (the driver) is known in advance. A purely blind approach, which does not exploit this information, seems suboptimal. On the other hand, adaptive beamforming algorithms require a DTD detector to be designed and thresholds to be tuned. Therefore, the second objective of the book is the development of a microphone array processing algorithm that combines the benefits of both approaches. In other words, we want to develop a method that (1) efficiently extracts the speech of one desired speaker from mixtures of multiple speakers, and (2) removes the need for a DTD and allows continuous adaptation, also during double-talk.

Moreover, an emphasis is placed on the evaluation of these methods with *real microphone recordings* involving simultaneous speech of the driver and the codriver, as opposed to computer-generated simulations. Experiments with background noise are carried out to assess the robustness of the considered methods in noisy conditions.

Let us set the limits of this book: Firstly, the attenuation of the background noise, which may be tackled with frequency-domain postfiltering techniques [14, 94], does not come into focus. Secondly, the design of a DTD will not be investigated. An approach that is well adapted for the automotive context can be found in [65]. For the sake of comparison with DTD-controlled beamformers, we will use an "informed" virtual DTD, based on the knowledge of the true target signal.

1.3 Outline of the Book

Chapter 2 is an introductory chapter which sets the formal framework on which the next chapters are based. It defines the notations, illustrates adaptive algorithms on simple examples, and defines performance measures. Then, Chaps. 3–9 can be divided into three parts:

- The first part (Chaps. 3 and 4) deals with LCMV beamforming. Chapter 3 introduces fundamental concepts in LCMV beamforming. Chapter 4 presents an "implicit" control scheme, as opposed to "explicit" double-talk detection. The proposed threshold-free adaptation control is a modification of the standard NLMS algorithm which mitigates the target signal cancelation problem.

- The second part (Chaps. 5–7) is dedicated to BSS. In Chap. 5, we consider the time-domain BSS method presented by Buchner et al. which exploits second-order statistics of the source signals [18, 20]. This method is based on the natural gradient and is limited to "square" systems with equally many sources and microphones. Introducing the concept of "partial separation," we propose a new approach to remove this restriction of the natural gradient. The Sylvester-based representation of the separation system allows a very concise derivation of Second-Order Statistics BSS (SOS-BSS) algorithms in the time-domain but cannot be directly implemented. Revisiting the natural gradient in the z-domain, we clarify this implementation issue in Chap. 6. Chapter 7 discusses the convergence and stability of SOS-BSS algorithms from the theoretical point of view.
- Chapters 8 and 9 constitute the last part of the book. Chapter 8 provides a detailed comparison of the two approaches. Chapter 9 examines existing and new combinations of SOS-BSS with beamforming. It also investigates these combinations as an acoustic front-end for a speech recognizer.

2

Source Separation as a Multichannel Linear Filtering Problem

The physical phenomena implied in multichannel speech processing include speech production at the vocal strings, sound propagation from the sound source to the microphone membrane, and analog-to-digital conversion of the signal. Instead of taking the complexity of these various phenomena into account, the acoustic signal processing algorithms studied in this book are based on the simplified model of a linear acoustic mixing.

This chapter explains this simplified model and is organized as follows: Section 2.1 describes the acoustic environment from the physical point of view and gives a mathematical formulation of the linear acoustic mixing model. In Sect. 2.2, the multichannel separation filters are presented for single and multiple outputs systems. The least-mean square (LMS) algorithm and a simple blind source separation (BSS) algorithm are briefly introduced to exemplify multichannel adaptive algorithms. The spatial response is introduced as a tool to interpret the separation filters spatially. In Sect. 2.3, we examine how the separation may be achieved and a lower bound on the length of the separation filters is derived. Finally, Sect. 2.4 defines the performance measures that will be used throughout the next chapters.

2.1 The Mixing Channels

For normal sound pressure in speech applications, the propagation medium (the air) and the transducer (the microphone) may be assumed to behave linearly. Hence, the emitted sound undergoes a linear transformation before reaching a given microphone. The filter that characterizes this transformation is completely described by its impulse response, which we denote by $h_{\mathbf{r},\boldsymbol{\theta}}$. This response is also called acoustic channel or room impulse response. The acoustic channel $h_{\mathbf{r},\boldsymbol{\theta}}$ depends on a set of parameters that may be divided into four groups:

- The position $\boldsymbol{\theta} \in \mathbb{R}^3$ where the *source signal* $s_{\boldsymbol{\theta}}(t)$ is emitted may be estimated from the input data, but it is more difficult to determine the orientation and radiation diagram of the source.
- The position $\mathbf{r} \in \mathbb{R}^3$ where the sound is received, i.e., the position[1] of the *microphone*, is known in most applications.
- The geometry of the *room* and the acoustic properties of its walls and objects which reflect the sound may hardly be modeled for real rooms and are unknown a priori in most applications.
- The propagation of sound in the *propagation medium* depends on parameters such as the temperature. Assuming a temperature of 20°C, the sound velocity equals $c = 344$ m s.

As an example, a room impulse response estimated in the car interior from the driver mouth to a microphone mounted in the rear-view mirror is shown in Fig. 2.1.

In general, the sound is produced in a region $\boldsymbol{\Gamma} \subset \mathbb{R}^3$ of the 3D space. In the noiseless case, the received signal $x(\mathbf{r}, t)$ is then obtained by integrating the convolution of the source signal $s_{\boldsymbol{\theta}}(t)$ with the response $h_{\mathbf{r},\boldsymbol{\theta}}$ over the region $\boldsymbol{\Gamma}$:

$$x(\mathbf{r}, t) = \int_{\boldsymbol{\Gamma}} h_{\mathbf{r},\boldsymbol{\theta}} * s_{\boldsymbol{\theta}} \, \mathrm{d}\boldsymbol{\theta}, \tag{2.1}$$

$$= \int_{\boldsymbol{\Gamma}} \int_0^{+\infty} h_{\mathbf{r},\boldsymbol{\theta}}(\tau) s_{\boldsymbol{\theta}}(t - \tau) \, \mathrm{d}\tau \, \mathrm{d}\boldsymbol{\theta}. \tag{2.2}$$

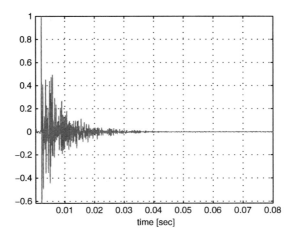

Fig. 2.1. Estimated impulse response between a close-talk microphone mounted on the artificial head on the driver seat and a microphone mounted on the four-element compact array mounted in the rear-view mirror

---

[1] The microphone orientation and sensitivity diagram may also be modeled. However these parameters are generally given from the manufacturer only within a certain tolerance margin.

When a human speaker or a loudspeaker is recorded by a distant microphone, the size of the region Γ is small relative to the source-microphone distance. Thus the source may be modeled as a point source, which is localized at a single spatial point θ_0. This is formally expressed by an excitation signal $s_{\theta_0}(t)$ as $s_\theta(t) = \delta(\theta - \theta_0)s_{\theta_0}(t)$. In this case, the space integral in (2.1) disappears and only the time integral remains:

$$x(\mathbf{r}, t) = h_{\mathbf{r}, \theta_0} * s_{\theta_0}, \tag{2.3}$$

$$= \int_0^{+\infty} h_{\mathbf{r}, \theta_0}(\tau)s_{\theta_0}(t - \tau)\,\mathrm{d}\tau. \tag{2.4}$$

If N point sources are present, we may write $s_\theta(t) = \sum_{n=1}^{N} \delta(\theta - \theta_n)s_{\theta_n}(t)$, which yields

$$x(\mathbf{r}, t) = \sum_{n=1}^{N} \int_0^{+\infty} h_{\mathbf{r}, \theta_n}(\tau)s_{\theta_n}(t - \tau)\,\mathrm{d}\tau. \tag{2.5}$$

In the remainder of this work we may omit the argument $\mathbf{r}$, the positions $\mathbf{r}_1, \ldots, \mathbf{r}_M$ of the M microphones being the ones we are interested in. In this case we simply note $x_m(t)$, $h_n(\tau)$, and $s_n(t)$ for $x(\mathbf{r}_m, t)$, $h_{\mathbf{r}, \theta_n}(\tau)$, and $s_{\theta_n}(t)$, respectively.

Since real room impulse responses decay at an exponential rate, they may be modeled by a finite impulse response (FIR) filter of length L_m. Let us denote the microphone signals sampled at time $t = p/f_s$ by $x(p)$, where $p \in \mathbb{Z}$ and where f_s denotes the sampling frequency. If we assume that the digitization is a linear operation, then the digitized microphone signal may be written from (2.5) as

$$x(p) = \sum_{n=1}^{N}(s_n * h_n)(p) = \sum_{n=1}^{N}\sum_{k=0}^{L_m} h_{n,k}s_n(p - k), \tag{2.6}$$

where $h_{n,k}$, $k = 0, \ldots, L_m$ denotes the digitized impulse response from the source $s_n(p)$ to the microphone $x(p)$. Equation (2.6) gives a formulation of the microphone signal as the linear mixing of the signal of interest and of the interferences (Fig. 2.2). In the case of several microphone signals $x_m(p)$, $m = 1, \ldots, M$, (2.6) turns into

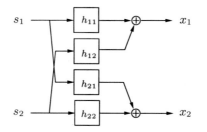

Fig. 2.2. Linear model for the acoustic mixing channels in the case $N = M = 2$

$$x_m(p) = \sum_{n=1}^{N} (h_{mn} * s_n)(p), \tag{2.7}$$

$$= \sum_{n=1}^{N} \sum_{k=0}^{L_{\mathrm{m}}} h_{mn,k} s_n(p - k). \tag{2.8}$$

Equation (2.8) models the acoustic mixing as a multiple-input multiple-output (MIMO) linear system. For later reference, we define the $L_{\mathrm{m}} \times 1$ vector $\mathbf{h}_{mn}$ as follows

$$\mathbf{h}_{mn} \triangleq \left(h_{mn,0}, \ldots, h_{mn,L_{\mathrm{m}}-1}\right)^{\mathrm{T}}. \tag{2.9}$$

It should be mentioned that (2.8) is a linear time-invariant *model* of the physical acoustic transmission channel. While this model describes the reality accurately in the case of loudspeaker-to-microphone transmission, it seems to be less accurate in the case of real mouth-to-microphone systems, even when the speakers are not moving [84].

In the remainder of this book, zero-mean signals are assumed.

2.2 The Separation Filters

This section introduces the separation filters that are algorithmically adjusted in order to extract the source of interest out of the acoustic mixing. We will present two kinds of separation systems: multiple-input single-output (MISO) and multiple-input multiple-output (MIMO) systems. We present MISO and MIMO systems separately because they are typically associated to distinct separation algorithms: While MISO systems are commonly used as interference cancelers in the context of informed beamforming, MIMO systems are the natural framework for blind source separation (BSS) techniques. The LMS algorithm and Van Gerven's SAD algorithm will be briefly presented to exemplify MISO and MIMO adaptive systems. Nevertheless, it should be mentioned that the concepts of MISO and MIMO systems are interchangeable in practice: The union of several MISO systems forms a MIMO system, and conversely the selection of a particular output in a MIMO system defines a MISO system. For simplicity, the separation filters are considered time-invariant unless stated otherwise.

2.2.1 Single Output Systems

According to the linear model in (2.8), the interferer signals are received through a linear system (the acoustic channels), and they may thus be canceled by linearly filtering the microphone signals. In single output systems, the output is obtained by filtering the M input signals individually and by taking the sum of the filtered signals. If we consider M FIR filters w_m of length L with coefficients $w_{m,l}, m = 1, \ldots, M; l = 0, \ldots, L-1$, this can be written as follows:

$$y(p) \triangleq \sum_{m=1}^{M} (x_m * w_m)(p), \tag{2.10}$$

$$= \sum_{m=1}^{M} \sum_{l=0}^{L-1} w_{m,l} x_m(p - l). \tag{2.11}$$

The convolution in (2.10) can be reformulated using a vector notation as follows:

$$y(p) = \sum_{m=1}^{M} \mathbf{w}_m^{\mathrm{T}} \mathbf{x}_m(p), \tag{2.12}$$

$$\text{with } \mathbf{w}_m \triangleq (w_{m,0}, \ldots, w_{m,L-1}),$$
$$\text{and } \mathbf{x}_m(p) \triangleq (x_m(p), \ldots, x_m(p - L + 1)).$$

In the context of MISO systems, it is usual to stack the input samples in the $ML \times 1$ vector $\mathbf{x}(p)$ and all filter coefficients into one $ML \times 1$ vector $\mathbf{w}$, which defines

$$\mathbf{x}(p) \triangleq \left[\mathbf{x}_1^{\mathrm{T}}(p), \ldots, \mathbf{x}_M^{\mathrm{T}}(p) \right]^{\mathrm{T}}, \quad \mathbf{w} \triangleq \left[\mathbf{w}_1^{\mathrm{T}}, \ldots, \mathbf{w}_M^{\mathrm{T}} \right]^{\mathrm{T}}. \tag{2.13}$$

Then we can compactly reformulate (2.12) as follows:

$$y(p) = \mathbf{w}^{\mathrm{T}} \mathbf{x}(p). \tag{2.14}$$

The squared L_2-norm of $\mathbf{w}$ is called its white-noise gain [88].

If we substitute the input signals $x_m(p)$ in (2.8), we can write $y(p)$ in terms of the source signals as follows:

$$y(p) = \sum_{m=1}^{M} (w_m * h_{mn} * s_n)(p) + \sum_{\substack{n'=1 \\ n' \neq n}}^{N} \sum_{m=1}^{M} (w_m * h_{mn'} * s_{n'})(p). \tag{2.15}$$

For a certain source of interest s_n, (2.15) decomposes $y(p)$ into the sum of the filtered desired source signal s_n and of the filtered interference signals $s_{n'}, n' \neq n$. Our objective is to find the separation filters w_m so that the second part of the sum (the filtered interfering source signals) vanishes. Suppose we define a cost function (or "criterion") J that is minimum if and only if this objective is reached. Then we could adjust the filter coefficients automatically using a minimization method such as e.g., the gradient descent. This concept is rather general and applies to most adaptive filtering algorithms.

The filter coefficients may be adjusted based on the statistics of the input or output signals. To describe these statistics, we use the expectation operator (or "ensemble average") $\mathbf{E}\{\}$ whose argument is a stochastic process. For the sake of simplicity, no notational distinction is made between the *realization*

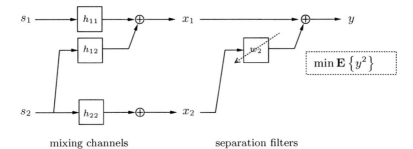

Fig. 2.3. Mixing/separation structure for Widrow's interference canceler for $M = 2$. In this context, the input $x_1(p)$ is referred to as the target reference signal. The input $x_2(p)$, which should not receive any contribution of the desired source, is referred to as the interference reference signal

$x(p)$ and the underlying stochastic process $X(p)$; we denote the expectation of the underlying stochastic process by $\mathbf{E}\{x(p)\}$.

For widely used adaptive filters, such as linearly constrained minimum variance (LCMV) beamformers or Widrow's interference canceler, the cost function is based on the output signal energy [92].

Example: Widrow's interference canceler and the LMS algorithm

In the following, we briefly sketch Widrow's interference canceler for two microphones ($M = 2$) and two sources ($N = 2$). The interference canceler is depicted in Fig. 2.3. We assume stationary input signals and define the cost function as the energy of the output signal, that is,

$$J_{\mathrm{LMS}} \triangleq \mathbf{E}\left\{y^2(p)\right\}. \tag{2.16}$$

To avoid the undesired solution $\mathbf{w} = \mathbf{0}$, the filter $\mathbf{w}_1$ is set to a unit impulse and only the filter $\mathbf{w}_2$ is adapted, as depicted in Fig. 2.3. The optimal solution which minimizes J_{LMS} may be computed iteratively using a gradient descent, as follows:

$$\mathbf{w}_2(n+1) \triangleq \mathbf{w}_2(n) - \mu \frac{\partial J_{\mathrm{LMS}}}{\partial \mathbf{w}_2}, \tag{2.17}$$

$$= \mathbf{w}_2(n) - 2\mu \mathbf{E}\left\{y(p)\mathbf{x}_2(p)\right\}, \tag{2.18}$$

where μ denotes the gradient descent step-size and n denotes the iteration index. From (2.18), it may be seen that the optimal filter coefficients also cancel the correlation $\mathbf{E}\{y(p)\mathbf{x}_2(p)\}$ between the interference input x_2 and the output y, in addition to minimizing the output energy. This implies certain restrictive conditions on the mixing channels for the convergence of (2.18) to the desired solution: The interference input x_2 should be free of any target signal, that is, the acoustic channel h_{21} should be zero. Otherwise the target signal will also be canceled at the output $y(p)$.

To obtain the online LMS algorithm, we replace the expectation with its instantaneous estimate. The cost function (2.16) becomes

$$J_{\text{online}}(p) \triangleq y^2(p). \tag{2.19}$$

If we consider $n = 0, \ldots, N_{\text{iter}} - 1$ iterations per input sample, the gradient descent may be written as

$$\mathbf{w}_2(n+1, p) \triangleq \mathbf{w}_2(n, p) - \mu \frac{\partial J_{\text{online}}(p)}{\partial \mathbf{w}_2}, \tag{2.20}$$

with $\mathbf{w}_2(0, p) = \mathbf{w}_2(N_{\text{iter}}, p - 1)$ as initialization for each p. When a single iteration for each new input sample is sufficient, the iteration index n in (2.20) may be dropped and one obtains the LMS algorithm

$$\mathbf{w}_2(p+1) = \mathbf{w}_2(p) - 2\mu y(p)\mathbf{x}_2(p). \tag{2.21}$$

The LMS algorithm and its application to speech separation will be studied in more detail in Chaps. 3 and 4.

2.2.2 Multiple Output Systems

The MISO separation architecture may be extended to N output signals with NM FIR filters $w_{nm}, n = 1, \ldots, N; m = 1, \ldots, M$ as follows:

$$y_n(p) \triangleq \sum_{m=1}^{M} (x_m * w_{nm})(p). \tag{2.22}$$

As in (2.12), we can rewrite (2.22) using a vector notation:

$$y_n(p) = \sum_{m=1}^{M} \mathbf{w}_{nm}^{\text{T}} \mathbf{x}_m(p), \tag{2.23}$$

with $\mathbf{w}_{nm} \triangleq (w_{nm,0}, \ldots, w_{nm,L-1})^{\text{T}}$. The overall mixing/separation system is represented in Fig. 2.4 for $N = M = 2$. In the context of MIMO systems, the filters $\mathbf{h}_{nn}$ and $\mathbf{w}_{nn}$ for $n = 1, \ldots, N$ will be referred to as *diagonal* filters.

MIMO systems are the standard framework for BSS techniques. In BSS, the sources are assumed to be mutually independent; hence, the separation may be achieved with a cost function $J(y_1, \ldots, y_N)$ that measures the dependence of the output signals.

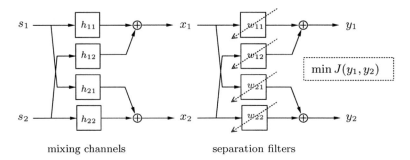

Fig. 2.4. MIMO structure in the case $N = M = 2$. The filter coefficients are adjusted to minimize a cost function J

Example of Van Gerven's symmetric adaptive decorrelation (SAD) algorithm

In the following, we briefly sketch a simple BSS algorithm which may be obtained by modifying the LMS algorithm, as proposed by Van Gerven [38]. We recall that the LMS algorithm decorrelates a signal estimate $y(p)$ from an interference reference $x_2(p)$. This leads to an undesired signal cancelation at the system output if the target signal $s_1(p)$ leaks into the interference reference $x_2(p)$, that is, whenever h_{21} is nonzero. Based on this observation, Van Gerven proposed replacing the interferer reference signal $\mathbf{x}_2(p)$ in (2.21) by the target-free interference estimate $y_2(p) = x_2(p) + \mathbf{w}_{21}^{\mathrm{T}}(p)\mathbf{x}_1(p)$. He then obtains two update rules for $\mathbf{w}_{12}(p)$ and $\mathbf{w}_{21}(p)$:

$$\begin{cases} \mathbf{w}_{12}(p+1) \triangleq \mathbf{w}_{12}(p) - \mu y_1(p)\mathbf{y}_2(p), \\ \mathbf{w}_{21}(p+1) \triangleq \mathbf{w}_{21}(p) - \mu y_2(p)\mathbf{y}_1(p). \end{cases} \tag{2.24}$$

Similarly to Widrow's interference canceler, the two diagonal filters $\mathbf{w}_{11}$ and $\mathbf{w}_{22}$ are set to unit impulses and their coefficients are fixed, as shown in Fig. 2.6 for $D = 0$. Even though (2.24) may overcome the limitations of the LMS algorithm (2.21) in case of target leakage (i.e., if h_{21} is nonzero), this simple algorithm is generally not able to separate speech sources in real acoustic environments [39]. In Chaps. 5–7, we will develop more robust BSS algorithms.

2.2.3 The Spatial Response

The spatial response is a tool to interpret the separation filters spatially. To introduce its formal definition, we consider a single source $s_{\boldsymbol{\theta}}(p)$ at position $\boldsymbol{\theta} \in \mathbb{R}^3$ which emits a unit impulse at time[2] $p = 0$ and denote the impulse

---

[2] The time at which the source $s_{\boldsymbol{\theta}}(p)$ emits a unit impulse is irrelevant, the impulse responses modeled in (2.26) may be shifted by a common, overall delay D. Hence we may replace (2.26) with $\mathbf{h}_{m,\boldsymbol{\theta}} = \boldsymbol{\delta}_{D+f_s\tau_{m,\boldsymbol{\theta}}}$. This overall delay D may be necessary to have causal filters if there exists m so that $\tau_{m,\boldsymbol{\theta}} < 0$.

response for the acoustic channel from $\boldsymbol{\theta}$ to the mth microphone by $h_{m,\boldsymbol{\theta}}$. Without loss of generality, we consider a MISO system with filters $\mathbf{w}_m$ for $m = 1, \ldots, M$. If only the source $s_{\boldsymbol{\theta}}(p)$ is present, its output can be written as

$$y(p) = \sum_{m=1}^{M} (\mathbf{w}_m * \mathbf{h}_{m,\boldsymbol{\theta}})(p). \tag{2.25}$$

The actual impulse responses $\mathbf{h}_{m,\boldsymbol{\theta}}$ are unknown in general. To interpret the coefficients $\mathbf{w}$ spatially, we may assume a simplified case where the acoustic propagation channels simply delay the source signals by $f_s \tau_{m,\boldsymbol{\theta}}$ samples (see footnote 1), that is

$$\mathbf{h}_{m,\boldsymbol{\theta}} = \boldsymbol{\delta}_{f_s \tau_{m,\boldsymbol{\theta}}}. \tag{2.26}$$

The delays $\tau_{m,\boldsymbol{\theta}}$ for $m = 1, \ldots, M$ may be computed from the source–microphone distances $\|\boldsymbol{\theta} - \mathbf{r}_m\|$. (Since the delays $f_s \tau_{m,\boldsymbol{\theta}}$ do not take integer values in general, fractional-delay filters are necessary to approximate $\boldsymbol{\delta}_{f_s \tau_{m,\boldsymbol{\theta}}}$ [64].) The spatial response for the position $\boldsymbol{\theta}$ is denoted by $\mathbf{g}(\boldsymbol{\theta})$ and is defined as the system output when only the source $s_{\boldsymbol{\theta}}(p)$ is present:

$$\mathbf{g}(\mathbf{w}, \boldsymbol{\theta}) \triangleq \sum_{m=1}^{M} \mathbf{w_m} * \boldsymbol{\delta}_{f_s \tau_{m,\boldsymbol{\theta}}}. \tag{2.27}$$

Case of the far- and free-field propagation model

If the source-microphone distances $\|\boldsymbol{\theta} - \mathbf{r}_m\|$ are large compared to the array aperture, the acoustic channels may be further simplified to the far- and free-field propagation model (see Appendix B for more details). In this case, the dependency in the 3D parameter $\boldsymbol{\theta}$ reduces to a dependency in the direction-of-arrival (DOA) θ, and for a uniform linear array (ULA) with interelement spacing Δ we have

$$\tau_{m,\theta} = (m-1)\frac{\Delta \sin(\theta)}{c}, \tag{2.28}$$

for all m. This simplification is widely used for compact microphone arrays. When the far- and free-field propagation model is assumed, we may denote the position parameter by θ instead of $\boldsymbol{\theta}$.

Representation in the DTFT domain

To formulate (2.27) using the DTFT,[3] we first define

---

[3] For any finite-length time-domain sequence $u(p)$, its DTFT is denoted by $U(\omega)$, where ω is a continuous angular frequency in the range $[-\pi, \pi]$, and is defined as follows:

$$U(\omega) = \sum_{p \in \mathbb{Z}} u(p)\, e^{-i\omega p}. \tag{2.29}$$

Note that $u(p)$ may be a signal or a filter.

$$\mathbf{W}(\omega) \triangleq (W_1(\omega), \dots, W_M(\omega))^{\mathrm{T}}, \qquad (2.30)$$

$$\mathbf{D}(\omega, \boldsymbol{\theta}) \triangleq (D_1(\omega, \boldsymbol{\theta}), \dots, D_M(\omega, \boldsymbol{\theta}))^{\mathrm{T}}, \qquad (2.31)$$

$$\text{with} D_m(\omega, \boldsymbol{\theta}) \triangleq e^{-\mathrm{i}\omega f_s \tau_{m,\theta}} \text{ for } m = 1, \dots, M. \qquad (2.32)$$

$\mathbf{D}(\omega, \boldsymbol{\theta})$ is called the *steering vector* for the position $\boldsymbol{\theta}$. The space-frequency response of the beamformer is given by

$$G_{\mathbf{w}}(\omega, \boldsymbol{\theta}) \triangleq \mathbf{W}^{\mathrm{T}}(\omega) \mathbf{D}(\omega, \boldsymbol{\theta}). \qquad (2.33)$$

If we assume a far- and free-field propagation, the squared magnitude of the space-frequency response may be represented in two dimensions (angle/frequency). The resulting figure is called the *beampattern* and is denoted by

$$\mathrm{BP}_{\mathbf{w}}(\omega, \theta) \triangleq |G_{\mathbf{W}}(\omega, \theta)|^2, \qquad (2.34)$$

$$= \left| \sum_{m=1}^{M} W_m(\omega)\, e^{-\mathrm{i}\omega \tau_{m,\theta}} \right|^2, \qquad (2.35)$$

where $\tau_{m,\theta}$ is given in (2.28).

To illustrate the beampattern, we consider a beamformer with $M = 4$ microphones and spacing $\Delta = 5\,\mathrm{cm}$. For this example, the filter coefficients are set to

$$\mathbf{w}_m = \frac{1}{M} \boldsymbol{\delta}_{f_s \tau_{m,-\theta_0}}, \qquad (2.36)$$

with $\theta_0 = 20°$. This beamformer is a delay-and-sum beamformer: The delays $\tau_{m,-\theta_0}$ synchronize the input signal for a certain DOA θ_0. After synchronization, the input signals are averaged, which enhance the signals coming from θ_0 and attenuates the signals coming from the other directions. The beampattern of this delay-and-sum beamformer is shown in Fig. 2.5.

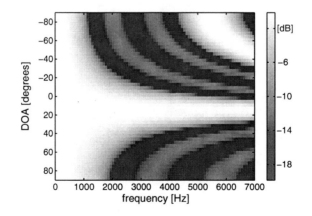

Fig. 2.5. Beampattern for the delay-and-sum beamformer. The steered DOA is $\theta_1 = 20°$. Other parameters: $M = 4, \Delta = 5\,\mathrm{cm}$

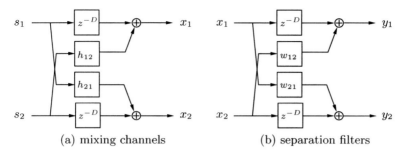

(a) mixing channels   (b) separation filters

Fig. 2.6. Linear mixing (resp. separation) structure in the case $N = M = 2$ with unit diagonal channels (resp. filters)

2.2.4 Particular Cases

Unit diagonal separation filters

A rather common simplification consists in constraining the diagonal filters to simple delays of D taps:

$$\mathbf{w}_{nn} = \boldsymbol{\delta}_D, \tag{2.37}$$

for $n = 1, \ldots, N$ and where D should be chosen in the range $0 \le D < L$. The resulting structure is depicted in Fig. 2.6b. This constraint may be applied to prevent the filter coefficients from converging to an undesired solution, as in the case of the interference canceler. It comes with a reduction of the number of degrees of freedom.

Instantaneous mixing

If we set $L_m = L = 1$ in (2.8) and (2.10), we obtain a so-called instantaneous mixing which does not involve any time delay between the source and the observed signals. In this case, each impulse response $h_{mn,k}, k = 0, \ldots, L_m - 1$ is reduced to the scalar factor $h_{mn,0}$. With such an excessive simplification, the mathematical model obviously loses its physical relevance. However, the mixing equation (2.8) may be rewritten very simply using a matrix notation as

$$\mathbf{x}(p) = \mathbf{H}\mathbf{s}(p), \tag{2.38}$$

with $\mathbf{x}(p) = (x_1(p), \ldots, x_M(p))^{\mathrm{T}}$, $\mathbf{s}(p) = (x_n(p), \ldots, s_N(p))^{\mathrm{T}}$, and

$$\mathbf{H} \triangleq \begin{bmatrix} h_{11,0} & \cdots & h_{1N,0} \\ \vdots & \ddots & \vdots \\ h_{M1,0} & \cdots & h_{MN,0} \end{bmatrix}. \tag{2.39}$$

Similarly, with $L = 1$, the separation filters reduce to separation scalar coefficients $w_{nm,0}$ that may be stacked into a single $N \times M$ matrix:

$$\mathbf{W} \triangleq \begin{bmatrix} w_{11,0} & \cdots & w_{1M,0} \\ \vdots & \ddots & \vdots \\ w_{N1,0} & \cdots & w_{NM,0} \end{bmatrix}. \tag{2.40}$$

The separation equation (2.22) can be then written as

$$\mathbf{y}(p) = \mathbf{W}\mathbf{x}(p), \tag{2.41}$$

with $\mathbf{y}(p) = (y_1(p), \ldots, y_N(p))^{\mathrm{T}}$. Instantaneous mixings may not be used directly to model acoustic mixtures, since time-delayed and multipath propagation always arise in reality. In spite of this, because of their simplicity, instantaneous mixings have been studied to a great extent in the field of BSS (see, e.g., [23] and the references therein).

2.3 Spatial Filtering vs. Spectral Filtering

The aim is now to find the filter coefficients $w_{nm,k}$ that cancel the interference signal. In the following, we examine how this may be done. Without loss of generality, we examine the case of a MISO system defined by its filters $\mathbf{w}_m$ for $m = 1, \ldots, M$ as in (2.12). The source of interest is $s_1(p)$, the other $N-1$ sources being considered as interferences. We consider the *source–output* MISO systems, described by the filters[4] $\mathbf{c}_n$ of length $L + L_{\mathrm{m}} - 1$:

$$\mathbf{c}_n \triangleq \sum_{m=1}^{M} \mathbf{w}_m * \mathbf{h}_{mn}. \tag{2.44}$$

To convolve the interference signals with the source–output filters, we introduce the $L + L_{\mathrm{m}} - 1 \times 1$ vectors

$$\mathbf{s}_n(p) \triangleq (s_n(p), \ldots, s_n(p - L - L_{\mathrm{m}} + 2)). \tag{2.45}$$

---

[4] In (2.44), we use the vector convolution. The convolution $\mathbf{c} = \mathbf{a} * \mathbf{b}$ of vectors $\mathbf{a}$ and $\mathbf{b}$ with respective lengths L_a and L_b is a vector $\mathbf{c}$ of length $L_c = L_a + L_b - 1$. This vector $\mathbf{c} = \mathbf{a} * \mathbf{b}$ is given by

$$\mathbf{c} \triangleq (c_0, \ldots, c_{L_a + L_b - 2})^{\mathrm{T}}, \tag{2.42}$$

$$\text{with } c_k \triangleq \sum_{p=\max\{0, k - L_b + 1\}}^{\min\{k, L_a - 1\}} a_p b_{k-p}. \tag{2.43}$$

We next stack the interference vector signals and filters into vectors of size $(N - 1)(L + L_m - 1) \times 1$:

$$\mathbf{c}_{\text{int}} \triangleq \left[\mathbf{c}_2^T, \ldots, \mathbf{c}_N^T \right]^T, \tag{2.46}$$

$$\mathbf{s}_{\text{int}}(p) \triangleq \left[\mathbf{s}_2^T(p), \ldots, \mathbf{s}_N^T(p) \right]^T. \tag{2.47}$$

Now the contribution of the interferences at the output, denoted $y_{\text{int}}(p)$, can be written as follows:

$$y_{\text{int}}(p) = \mathbf{c}_{\text{int}}^T \mathbf{s}_{\text{int}}(p). \tag{2.48}$$

The interferences are canceled by the filters $\mathbf{w}_n$ if the average power of $y_{\text{int}}(p)$ is zero. According to (2.48) and assuming stationary signals, this is achieved if and only if

$$\mathbf{c}_{\text{int}}^T \mathbf{R}_{\text{int}} \mathbf{c}_{\text{int}} = 0, \tag{2.49}$$

where $\mathbf{R}_{\text{int}}$ denotes the correlation matrix of $\mathbf{s}_{\text{int}}$ and is defined as

$$\mathbf{R}_{\text{int}} \triangleq \mathbf{E}\left\{ \mathbf{s}_{\text{int}}(p)\mathbf{s}_{\text{int}}^T(p) \right\}. \tag{2.50}$$

Let us assume that $\mathbf{R}_{\text{int}} \neq \mathbf{0}$ (the interferer is active). We may distinguish two particular solutions to (2.49).

Firstly, (2.49) is solved if $\mathbf{c}_{\text{int}} = \mathbf{0}$, which defines $(N - 1)(L + L_m - 1)$ constraints. The solution $\mathbf{c}_{\text{int}} = \mathbf{0}$ yields the advantage to be independent of the interference correlation matrix $\mathbf{R}_{\text{int}}$ and depends only on the acoustic mixing channels. Hence, we say that this solution achieves the separation by filtering the input signals *spatially*. This may be attained if the separation filters are long enough, as discussed in Sect. 2.3.1.

Secondly, (2.49) may be solved if $\mathbf{c}_{\text{int}}$ belongs to the null space of $\mathbf{R}_{\text{int}}$. This solution depends on the mixing channels but also on the power spectrum of the interference signals. Then we may say that the separation is achieved by filtering the input signals *spectrally*. In the case of nonstationary signals, this solution has the drawback to be time-varying, making it more difficult to be temporally tracked. Moreover, this yields a distortion of the desired signal if its spectrum overlaps with that of the interference signals.

Neither spatial nor spectral separation may be completely achieved. For example, the separation filters may be too short for spatial separation and the spectrum of the interferers may be too wide for spectral separation (e.g., in the case of white interference signals). Even if the separation filters are long enough, the adapted filter coefficients may depend on the excitation source signals, for example, if the spectrum of the excitation signal is not sufficiently wide. In practice, the suppression of the interferer signals relies on both spatial filtering (which sets $\mathbf{c}_{\text{int}}$ to $\mathbf{0}$) and spectral filtering (which sets $\mathbf{c}_{\text{int}}$ in the null space of $\mathbf{R}_{\text{int}}$). As we will see in Sect. 2.3.2, the case of instantaneous mixings is very special: If $L = 1$, spectral separation automatically achieves spatial separation, i.e., the optimal filters are always independent of the source signal.

2.3.1 Minimum Filter Length for Spatial Separation

We want to find the separating filter coefficients $w_{m,k}$ that achieve

$$\mathbf{c}_{\text{int}} = \mathbf{0}, \tag{2.51}$$

where $\mathbf{c}_{\text{int}}$ is defined in (2.44) and (2.46). In the following, we derive the minimum filter length for identifying these *interference-independent* filter coefficients. A related analysis may be found in [51].

Equation (2.51) defines a linear system. First, we represent this linear system under matrix form. For that, we define the $L + L_{\text{m}} - 1 \times L$ matrices $\mathbf{H}_{mn}$ as

$$\mathbf{H}_{mn} \triangleq \begin{pmatrix} h_{mn,L_{\text{m}}-1} & 0 & \cdots & 0 \\ \vdots & h_{mn,L_{\text{m}}-1} & & \vdots \\ h_{mn,0} & \vdots & \ddots & 0 \\ 0 & h_{mn,0} & & h_{mn,L_{\text{m}}-1} \\ \vdots & & \ddots & \vdots \\ 0 & \cdots & 0 & h_{mn,0} \end{pmatrix}, \tag{2.52}$$

for $m = 1, \ldots, M$ and $n = 1, \ldots, N$. By stacking matrices $\mathbf{H}_{mn}$ together, we obtain the $(N-1)(L + L_{\text{m}} - 1) \times ML$ interference mixing matrix:

$$\mathbf{H}_{\text{int}} \triangleq \begin{bmatrix} \mathbf{H}_{12} & \cdots & \mathbf{H}_{M2} \\ \vdots & & \vdots \\ \mathbf{H}_{1N} & \cdots & \mathbf{H}_{MN} \end{bmatrix}. \tag{2.53}$$

Now we can write (2.51) as follows:

$$\mathbf{H}_{\text{int}}\mathbf{w} = \mathbf{0}. \tag{2.54}$$

Depending on the number of linearly independent rows in $\mathbf{H}_{\text{int}}$, (that is, the row-rank of $\mathbf{H}_{\text{int}}$), (2.54) may have a solution or not. A plausible and common assumption is that the acoustic channels h_{mn} do not share common zeros in the frequency domain, which implies that the matrix $\mathbf{H}_{\text{int}}$ has full row-rank (see [51] and the references therein). Then (2.54) sets $(N-1)(L+L_{\text{m}}-1)$ linearly independent constraints on the ML separation filter coefficients $w_{m,l}$. To avoid the trivial zero solution $w_{m,l} = 0$ for all m, l, an additional constraint, e.g., $w_{1,0} = 1$ is required, leaving $ML - 1$ free parameters. The constraint in (2.51) can be fulfilled only if the number of degrees of freedom is not less than the number of constraints, i.e., only if $ML - 1 \geq (N-1)(L + L_{\text{m}} - 1)$. This yields the following lower bound on the length L of the separation filters:

$$L \geq \left\lceil \frac{(L_{\text{m}} - 1)(N - 1) + 1}{M - N + 1} \right\rceil, \tag{2.55}$$

where $\lceil x \rceil$ is the smallest integer larger than $x \in \mathbb{R}$. According to (2.55), in the case $L_{\text{m}} = 1$, spatial separation can be achieved with $L = 1$ provided that

$M \geq N$. For $L_{\mathrm{m}} > 1$ and $M > N$, spatial separation may be obtained with separation filters that are shorter than the mixing channels. In the square case ($M = N$), the spatial separation is attainable if $L \geq (L_{\mathrm{m}} - 1)(N - 1) + 1$. If in addition $N = 2$ then spatial separation is achievable if $L = L_{\mathrm{m}}$.

2.3.2 Particular Cases

Unit diagonal mixing/separation filters

The constraint in (2.37) comes with a reduction of the number of degrees of freedom, which needs to be taken into account for deriving the lower bound on L. Again, we formulate the separation constraints under matrix form for a MISO separation system with M filters $\mathbf{w}_m$, $m = 1, \ldots, M$ and $\mathbf{w}_1 = \boldsymbol{\delta}_D$. Using the matrices $\mathbf{H}_{mn}$ introduced in (2.52), we first define

$$\mathbf{H}'_{\mathrm{int}} \triangleq \begin{bmatrix} \mathbf{H}_{22} & \cdots & \mathbf{H}_{M2} \\ \vdots & & \vdots \\ \mathbf{H}_{2N} & \cdots & \mathbf{H}_{MN} \end{bmatrix}, \quad \mathbf{w}' \triangleq \begin{bmatrix} \mathbf{w}_2 \\ \vdots \\ \mathbf{w}_M \end{bmatrix}, \quad \mathbf{h}' \triangleq -\begin{bmatrix} \mathbf{H}_{12} \\ \vdots \\ \mathbf{H}_{1M} \end{bmatrix} \boldsymbol{\delta}_D. \quad (2.56)$$

Now the separation constraint may be written as follows:

$$\mathbf{H}'_{\mathrm{int}} \mathbf{w}' = \mathbf{h}'. \quad (2.57)$$

First we consider the case where $\mathbf{H}'_{\mathrm{int}}$ has full row-rank. Then (2.57) defines $(N - 1)(L + L_{\mathrm{m}} - 1)$ linear constraints. Since there are $(M - 1)L$ degrees of freedom, an interference-independent separation may be achieved only if

$$(M - N)L \geq \lceil (L_{\mathrm{m}} - 1)(N - 1) \rceil, \quad (2.58)$$

$$\Leftrightarrow L \geq \left\lceil \frac{(L_{\mathrm{m}} - 1)(N - 1)}{M - N} \right\rceil \quad \text{if } M > N. \quad (2.59)$$

Equation (2.58) shows that the $(N - 1)(L + L_{\mathrm{m}} - 1)$ constraints in (2.51) cannot be fulfilled if $M = N$ unless $L_{\mathrm{m}} = 1$. This means that no interference-independent separating solution can be identified.

Now we give an example where $\mathbf{H}'_{\mathrm{int}}$ does not have full row-rank. Consider $M = N = 2$ and $\mathbf{h}_{22} = \mathbf{w}_1 = \boldsymbol{\delta}_D$, that is, the diagonal channel $\mathbf{h}_{22}$ has a unit response as depicted in Fig. 2.6a. If the delay D is zero, this may be considered as a physical model in situations where the room acoustics is not very reverberant and where the source s_2 is placed close to the microphone x_2 for $n = 1, 2$ [38]. Then a separating solution is given by

$$\mathbf{w}_2 = -\mathbf{h}_{12}. \quad (2.60)$$

Instantaneous mixing

As may be seen from (2.55) and (2.58), instantaneous mixings yield specific properties: If $L_m = 1$, an interference-independent separating solution can be always identified, even with unit diagonal filters as in (2.37).

2.4 Performance Measures

This section introduces the performance measures that are used as comparison criteria in the remainder of this work, in particular the signal-to-interference ratio (SIR) improvement. Since the performance measures are defined in terms of *improvement* with respect to the signal quality at the microphone input, it is necessary to chose an input reference. This choice differs depending on the type of microphone array: In the case of a compact beamformer, no input signal has a larger SIR than another a priori, hence the reference is obtained by averaging over the M input microphones. This is formulated in Sect. 2.4.1. By contrast, in the case of a distributed microphone array, the input signal x_1 is a priori known to have a large SIR, hence the reference is obtained on this particular input signal as formulated in Sect. 2.4.2.

The observed signals $x_m(p)$ may be decomposed as the sum of the contributions of the desired source and of those of the interferers:

$$x_m(p) = x_{\text{sig},m}(p) + x_{\text{int},m}(p). \tag{2.61}$$

Similarly, the output signal may be decomposed as follows:

$$y(p) = y_{\text{sig}}(p) + y_{\text{int}}(p), \tag{2.62}$$

$$\text{with } y_{\text{sig}}(p) = \sum_{m=1}^{M} (w_m * x_{\text{sig},m})(p), \tag{2.63}$$

$$\text{and } y_{\text{int}}(p) = \sum_{m=1}^{M} (w_m * x_{\text{int},m})(p). \tag{2.64}$$

Evaluating the performance measures requires signal powers to be estimated. For the definitions below, we use a generic estimate of the statistical expectation denoted by $\widehat{\mathbf{E}}\{\}$, whose actual implementation depends on the processing mode: In batch mode, we may use

$$\widehat{\mathbf{E}}\{f(\mathbf{A}(p))\} \triangleq \frac{1}{T} \sum_{p=1}^{T} f(\mathbf{A}(p)), \tag{2.65}$$

where T denotes the signal length (in samples). (The function f and the matrix $\mathbf{A}$ are placeholders that should be replaced by the variables of interest.) In online mode, the instantaneous estimate may be used:

$$\widehat{\mathbf{E}}\{f(\mathbf{A}(p))\} \triangleq f(\mathbf{A}(p)). \tag{2.66}$$

2.4.1 Compact Microphone Array

For compact microphone arrays, the following power ratios are introduced:

- the reduction of the target signal level, denoted by SR, defined as the ratio of the desired signal power at the sensors averaged over the M sensors signals and the desired signal power at the output:

$$SR(p) \triangleq \frac{\sum_{m=1}^{M} \widehat{\mathbf{E}}\left\{x_{\text{sig},m}^2(p)\right\}/M}{\widehat{\mathbf{E}}\left\{y_{\text{sig}}^2(p)\right\}}, \tag{2.67}$$

- the reduction of the interference signal level, IR, defined as the ratio of the interference signal power at the sensors averaged over the M sensors, and the interference signal power at the output:

$$IR(p) \triangleq \frac{\sum_{m=1}^{M} \widehat{\mathbf{E}}\left\{x_{\text{int},m}^2(p)\right\}/M}{\widehat{\mathbf{E}}\left\{y_{\text{int}}^2(p)\right\}}, \tag{2.68}$$

The signal-to-interference ratio improvement, SIR_{imp}, is defined as

$$SIR_{\text{imp}}(p) \triangleq \frac{IR(p)}{SR(p)}. \tag{2.69}$$

2.4.2 Distributed Microphone Array

For distributed microphone arrays, the input reference is taken at the sensor $x_1(p)$, which yields:

- the reduction of the target signal level, denoted by SR^d, defined as the ratio of the desired signal power at the sensor $x_1(p)$ and the desired signal power at the output:

$$SR^d(p) \triangleq \frac{\widehat{\mathbf{E}}\left\{x_{\text{sig},1}^2(p)\right\}}{\widehat{\mathbf{E}}\left\{y_{\text{sig}}^2(p)\right\}}, \tag{2.70}$$

- the reduction of the interference signal level, IR^d,

$$IR^d(p) \triangleq \frac{\widehat{\mathbf{E}}\left\{x_{\text{int},1}^2(p)\right\}}{\widehat{\mathbf{E}}\left\{y_{\text{int}}^2(p)\right\}}. \tag{2.71}$$

The signal-to-interference ratio improvement, SIR_{imp}^d, is defined as

$$SIR_{\text{imp}}^d(p) \triangleq \frac{IR^d(p)}{SR^d(p)}. \tag{2.72}$$

Table 2.1. Definition of the start-up performance and of the performance after initial convergence for the compact array and the distributed array. The subscript $[t_0, t_1]$ indicates an average of the SIR improvement from time $p = t_0$ to time $p = t_1$ (in seconds)

Compact array	Distributed array
$Q_{[0,3]} \triangleq IR_{[0,3]}/SR_{[0,3]},$	$Q^d_{[0,3]} \triangleq IR^d_{[0,3]}/SR^d_{[0,3]},$
$Q_{[3,10]} \triangleq IR_{[3,10]}/SR_{[3,10]}$	$Q^d_{[3,10]} \triangleq IR^d_{[3,10]}/SR^d_{[3,10]}$

$$IR_{[t_0,t_1]} \triangleq \frac{\dfrac{1}{M}\displaystyle\sum_{m=1}^{M}\sum_{p=t_0 f_s}^{t_1 f_s} x^2_{\text{int},m}(p)}{\displaystyle\sum_{p=t_0 f_s}^{t_1 f_s} y^2_{\text{int}}(p)} \qquad\qquad IR^d_{[t_0,t_1]} \triangleq \frac{\displaystyle\sum_{p=t_0 f_s}^{t_1 f_s} x^2_{\text{int},1}(p)}{\displaystyle\sum_{p=t_0 f_s}^{t_1 f_s} y^2_{\text{int}}(p)}$$

$$SR_{[t_0,t_1]} \triangleq \frac{\dfrac{1}{M}\displaystyle\sum_{m=1}^{M}\sum_{p=t_0 f_s}^{t_1 f_s} x^2_{\text{sig},m}(p)}{\displaystyle\sum_{p=t_0 f_s}^{t_1 f_s} y^2_{\text{sig}}(p)} \qquad\qquad SR^d_{[t_0,t_1]} \triangleq \frac{\displaystyle\sum_{p=t_0 f_s}^{t_1 f_s} x^2_{\text{sig},1}(p)}{\displaystyle\sum_{p=t_0 f_s}^{t_1 f_s} y^2_{\text{sig}}(p)}$$

2.4.3 Start-Up Performance and Performance after Initial Convergence

It seems difficult to compare different adaptive algorithms fairly: In particular, the step-size parameters may significantly influence the separation performance. To obtain an objective performance measure, two quantities are considered. First, we average the SIR improvement over the first three seconds, which gives $Q_{[0,3]}$ as defined in Table 2.1. The value $Q_{[0,3]}$ is used as an approximate measure of the speed of convergence during the initial convergence phase. Second, the average over the following seven seconds, as defined by $Q_{[3,10]}$ in Table 2.1, is considered. $Q_{[3,10]}$ gives an approximate measure of the performance after the initial convergence. It should be noted that the input signals also contain sensor noise for real recordings. Moreover, the averages include the silence periods. Therefore, the performance measures presented here are only *approximations* of the SIR improvement. That is why we prefer *not* denoting them by SIR_{imp} but by $Q_{[t_0,t_1]}$.

2.5 Summary and Conclusion

This chapter set the formal framework on which the next chapters are based. It modeled the acoustic environment carrying the source signal to the observed microphone signals as a MIMO linear system. The separation filters have been introduced as MISO and MIMO systems, which are algorithmically adapted to the input signals by minimizing a mathematically defined cost function. The LMS and the SAD algorithms, two exemplary algorithms that will be further developed in the next chapters, illustrated adaptive algorithms.

The spatial response, a tool to interpret the separation filters spatially, has been defined. We have seen that the cancelation of the interference signal may be achieved by filtering the input signal spatially or spectrally, depending on the number of degrees of freedom and on the constraints that are set on the separation filters. At last, we defined the performance measures that will be used in the following chapters to evaluate the algorithms under scope.

3

Linearly Constrained Minimum Variance Beamforming

The concept of "beamforming" refers to multichannel signal processing techniques that enhance the acoustic signals coming from a particular a priori known position, while reducing the signals coming from other directions. A number of beamforming techniques exist, a review of which may be found in [88]. In this chapter, we introduce linearly constrained minimum variance (LCMV) beamformers, which are widely used in acoustic array processing. The class of the LCMV beamformers is general enough to form a common framework to design beamforming algorithms for various physical setups.

Section 3.1 defines the LCMV beamforming principle. Section 3.2 provides the generalized sidelobe canceler (GSC) as an alternative formulation of the LCMV beamformer which can be implemented more efficiently. Section 3.3 shows how to apply the GSC to distributed and compact microphone arrays. Section 3.4 discusses the practical limitations of LCMV beamforming in reverberant environments.

3.1 LCMV Beamforming

Formal definition

Let us briefly recall the notations from Chap. 2: We consider a time-varying MISO system with filter coefficients $\mathbf{w}(p)$ and output $y(p) = \mathbf{w}^{\mathrm{T}}(p)\mathbf{x}(p)$ (see Sect. 2.2.1). In LCMV beamforming, the filter coefficients are adjusted based on the statistics of the output signals. To describe these statistics, we use the expectation operator $\mathbf{E}\{\}$ (or "ensemble average") whose argument is a stochastic process.[1]

---

[1] For the sake of simplicity, no notational distinction is made between the *realization* $x(p)$ and the underlying stochastic process $X(p)$. Hence, we denote the expectation of the underlying stochastic process by $\mathbf{E}\{x(p)\}$.

In LCMV beamforming, the cost function is the output signal variance. Since zero-mean signals are assumed, the cost function may be defined as the output signal power at time p, that is,

$$J(p) \triangleq \mathbf{E}\left\{y^2(p)\right\}. \tag{3.1}$$

Using the input correlation matrix

$$\mathbf{R}_{\mathbf{xx}}(p) \triangleq \mathbf{E}\left\{\mathbf{x}(p)\mathbf{x}^{\mathrm{T}}(p)\right\}, \tag{3.2}$$

we can rewrite $J(p)$ in (3.1) as a function of $\mathbf{w}(p)$:

$$J(p) = \mathbf{w}^{\mathrm{T}}(p)\mathbf{R}_{\mathbf{xx}}(p)\mathbf{w}(p). \tag{3.3}$$

Now, minimizing $J(p)$ may lead to $\mathbf{w}(p) = \mathbf{0}$ and $y(p) = 0$ for all p. In LCMV beamforming, this is prevented by constraining the filter coefficients linearly. For example, a simple linear constraint is that of Widrow's interference canceler where the filter $\mathbf{w}_1(p)$ is constrained to a unit impulse:

$$\mathbf{w}_1(p) = \boldsymbol{\delta}_0. \tag{3.4}$$

This constraint has dimension L. More generally, a linear constraint of dimension C may be formulated with a $ML \times C$ *constraint matrix* $\mathbf{C}$ and a $C \times 1$ *response vector* $\mathbf{c}$ as

$$\mathbf{C}^{\mathrm{T}}\mathbf{w}(p) = \mathbf{c}. \tag{3.5}$$

We find the constraint in (3.4) by setting

$$\mathbf{C}^{\mathrm{T}} = [\mathbf{I}_{L \times L}\ \mathbf{0}_{L \times L}], \qquad \mathbf{c} = \boldsymbol{\delta}_0. \tag{3.6}$$

Note that time-varying constraints may also be considered using a time-varying constraint matrix $\mathbf{C}(p)$ and a time-varying response vector $\mathbf{c}(p)$. For the sake of simplicity, we bound the presentation to time-invariant constraints. To summarize, LCMV beamforming consists in adjusting the filter coefficients according to the following constrained criterion:

$$\min_{\mathbf{w}(p)} \mathbf{E}\left\{y^2(p)\right\} \ \text{s.t.}\ \mathbf{C}^{\mathrm{T}}\mathbf{w}(p) = \mathbf{c}. \tag{3.7}$$

Constraining the spatial response

The motivation behind the linear constraint in (3.5) also comes from the fact that the spatial response of the beamformer, $\mathbf{g}(\mathbf{w}, \boldsymbol{\theta})$, is a linear function of $\mathbf{w}$, as shown in Sect. 2.2.3:

$$\mathbf{g}(\mathbf{w}, \boldsymbol{\theta}) \triangleq \sum_{m=1}^{M} \mathbf{w_m} * \boldsymbol{\delta}_{\tau_{m,\boldsymbol{\theta}}f_s}. \tag{2.27}$$

To obtain a given spatial response $\mathbf{g}_0$ for a particular position $\boldsymbol{\theta}_0$, one may set the constraint

$$\mathbf{g}(\mathbf{w}(p), \boldsymbol{\theta}_0) = \mathbf{g}_0 \tag{3.8}$$

for all p. Since $\mathbf{g}(\mathbf{w}, \boldsymbol{\theta}_0)$ in (2.27) depends linearly on the filter coefficients $\mathbf{w}_m(p)$, the constraint in (3.8) may be formulated using a certain matrix $\mathbf{C}$ as in (3.5). Typically, LCMV beamformers are designed to maintain a unit spatial response in the direction of the desired source known a priori. This should reduce the contribution of the interfering sources while keeping the desired source.

General solution to the LCMV optimization problem

In the following, we derive the solution to the LCMV optimization problem in (3.7) using the Lagrange multiplier method. For the sake of brevity and readability, we omit the time index p in the notations. The Lagrange cost function associated to (3.7) is given by

$$\mathcal{L}(\mathbf{w}, \boldsymbol{\lambda}) \triangleq \frac{1}{2}\mathbf{w}^{\mathrm{T}}\mathbf{R}_{\mathbf{xx}}\mathbf{w} + \boldsymbol{\lambda}^{\mathrm{T}}\left(\mathbf{C}^{\mathrm{T}}\mathbf{w} - \mathbf{c}\right). \tag{3.9}$$

The Lagrange multiplier $\boldsymbol{\lambda}$ is a $C \times 1$ vector. Setting the gradient of $\mathcal{L}(\mathbf{w}, \boldsymbol{\lambda})$ to zero, we obtain

$$\frac{\partial \mathcal{L}(\mathbf{w}, \boldsymbol{\lambda})}{\partial \mathbf{w}} = \mathbf{w}^{\mathrm{T}}\mathbf{R}_{\mathbf{xx}} + \boldsymbol{\lambda}^{\mathrm{T}}\mathbf{C} = \mathbf{0}, \tag{3.10}$$

and assuming that the correlation matrix $\mathbf{R}_{\mathbf{xx}}$ is nonsingular, we may write

$$\mathbf{w}^{\mathrm{T}} = -\boldsymbol{\lambda}^{\mathrm{T}}\mathbf{C}\mathbf{R}_{\mathbf{xx}}^{-1}. \tag{3.11}$$

Using the constraint $\mathbf{C}^{\mathrm{T}}\mathbf{w} = \mathbf{c}$, we transpose and multiply (3.11) with $\mathbf{C}^{\mathrm{T}}$ from the left to obtain

$$\boldsymbol{\lambda} = -\left(\mathbf{C}\mathbf{R}_{\mathbf{xx}}^{-\mathrm{T}}\mathbf{C}^{\mathrm{T}}\right)^{-1}\mathbf{c}. \tag{3.12}$$

Substituting $\boldsymbol{\lambda}$ in (3.12) into (3.11), we obtain the time-domain LCMV beamformer

$$\mathbf{w} = \mathbf{R}_{\mathbf{xx}}^{-\mathrm{T}}\mathbf{C}^{\mathrm{T}}\left(\mathbf{C}^{\mathrm{T}}\mathbf{R}_{\mathbf{xx}}^{-\mathrm{T}}\mathbf{C}^{\mathrm{T}}\right)^{-1}\mathbf{c}. \tag{3.13}$$

The computation of (3.13) involves the inversion of $\mathbf{R}_{\mathbf{xx}}$. In practice, estimates for $\mathbf{R}_{\mathbf{xx}}$ may be badly conditioned especially for large filter lengths L and colored input signals (such as speech). In addition, inverting $\mathbf{R}_{\mathbf{xx}}$ is computationally expensive. A wideband solution may be obtained by minimizing $\mathcal{L}(\mathbf{w}, \boldsymbol{\lambda})$ with an iterative gradient descent that avoids the matrix inversion [36]. It is much more efficient, however, to transform the *constrained* minimization problem into an *unconstrained* one. This is the principle of the generalized sidelobe canceler (GSC) [42].

3.2 From LCMV to Generalized Sidelobe Canceler (GSC)

Griffith and Jim introduced the generalized sidelobe canceler (GSC) as an alternative formulation of the LCMV problem which transforms the constrained minimization problem (3.7) into an unconstrained one [42].

Fixed beamformer and blocking matrix

The GSC is based on a decomposition of vector $\mathbf{w}(p)$ into two orthogonal components $\mathbf{w}_0$ and $\mathbf{v}$ with $\mathbf{w}(p) = \mathbf{w}_0 + \mathbf{v}(p)$. The first component $\mathbf{w}_0$ is fixed. It is chosen so that it satisfies the constraint

$$\mathbf{C}^{\mathrm{T}}\mathbf{w}_0 = \mathbf{c}. \tag{3.14}$$

This component is often (but not necessarily) set to $\mathbf{w}_0 = \mathbf{C}\left(\mathbf{C}^{\mathrm{T}}\mathbf{C}\right)^{-1}\mathbf{c}$. Such a choice minimizes its L_2-norm $\mathbf{w}_0^{\mathrm{T}}\mathbf{w}_0$, which is also called white-noise gain [88]. Because this component is not adapted, it is termed *fixed beamformer*. We define $x_0(p)$ as the output of the fixed beamformer, i.e., $x_0(p) = \mathbf{w}_0^{\mathrm{T}}\mathbf{x}(p)$.

The second component, $\mathbf{v}(p)$, is adapted but must belong to the subspace $\mathcal{V}$ of the filters that are orthogonal to the constraint, that is, $\mathcal{V} = \{\mathbf{v} \in \mathbb{R}^{ML \times 1} \text{ s.t. } \mathbf{C}^{\mathrm{T}}\mathbf{v} = \mathbf{0}_{C \times 1}\}$. Let $\mathbf{B}$ be an $ML \times M'L$ matrix whose $M'L$ columns span the subspace $\mathcal{V}$, i.e., so that

$$\mathrm{rank}(\mathbf{B}) = ML - C \quad \text{and} \quad \mathbf{C}^{\mathrm{T}}\mathbf{B} = \mathbf{0}_{C \times M'L}. \tag{3.15}$$

The matrix $\mathbf{B}$ is called the blocking matrix. Note that $\mathbf{B}$ is not uniquely determined by the equation $\mathbf{C}^{\mathrm{T}}\mathbf{B} = \mathbf{0}_{C \times M'L}$, several implementations of the blocking matrix are possible.

Adaptive interference canceler

For any $M'L \times 1$ vector $\mathbf{a}(p)$, the vector $\mathbf{v}(p) = \mathbf{B}\mathbf{a}(p)$ belongs to $\mathcal{V}$, hence the filter coefficients $\mathbf{w}(p) = \mathbf{w}_0 + \mathbf{B}\mathbf{a}(p)$ always fulfill the constraint $\mathbf{C}^{\mathrm{T}}\mathbf{w}(p) = \mathbf{c}$. Therefore, the constrained minimization (3.7) may be rewritten without constraint as

$$\min_{\mathbf{a}(p)} \left(\mathbf{w}_0 + \mathbf{B}\mathbf{a}(p)\right)^{\mathrm{T}} \mathbf{R}_{\mathbf{xx}}(p) \left(\mathbf{w}_0 + \mathbf{B}\mathbf{a}(p)\right). \tag{3.16}$$

Minimization affects only $\mathbf{a}(p)$, which is called the adaptive interference canceler. In GSC beamformers, the computation of the output involves the following variables:

$$x_0(p) \triangleq \mathbf{w}_0^{\mathrm{T}}\mathbf{x}(p), \tag{3.17}$$
$$\mathbf{x}_B(p) \triangleq \mathbf{B}^{\mathrm{T}}\mathbf{x}(p), \tag{3.18}$$
$$y(p) = x_0(p) + \mathbf{a}^{\mathrm{T}}(p)\mathbf{x}_B(p). \tag{3.19}$$

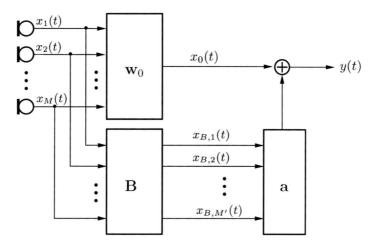

Fig. 3.1. Generalized Sidelobe Canceler

The Generalized Sidelobe Canceler structure is shown in Fig. 3.1. Since $x_0(p)$ should provide an enhanced target signal, it is referred to as the target signal reference. By contrast, $\mathbf{x}_B(p)$ is called the interference reference.

The output signal power is minimized with respect to $\mathbf{a}(p)$. The gradient of the cost function is

$$\frac{\partial J(p)}{\partial \mathbf{a}(p)} = 2\mathbf{E}\left\{x_0(p)\mathbf{x}_B(p)\right\} + 2\mathbf{E}\left\{\mathbf{x}_B(p)\mathbf{x}_B^{\mathrm{T}}(p)\right\}\mathbf{a}(p), \qquad (3.20)$$

$$= 2\mathbf{E}\left\{\mathbf{x}_B(p)y(p)\right\}. \qquad (3.21)$$

Setting $\mathbf{R}_{\mathbf{x}_B\mathbf{x}_B}(p) = \mathbf{E}\left\{\mathbf{x}_B(p)\mathbf{x}_B^{\mathrm{T}}(p)\right\}$ and the gradient in (3.20) to $\mathbf{0}$, we obtain the Wiener solution:

$$\mathbf{a}_{\mathrm{opt}}(p) = -\mathbf{R}_{\mathbf{x}_B\mathbf{x}_B}^{-1}(p)\mathbf{E}\left\{x_0(p)\mathbf{x}_B(p)\right\}. \qquad (3.22)$$

In addition to minimizing the output signal power, the Wiener solution (3.22) also decorrelates the input of the interference canceler $\mathbf{x}_B$ and the beamformer output y. This can be seen from the gradient expression in (3.21).

3.3 Constraints for Compact and Distributed Setups

3.3.1 Constraint for Compact Microphone Array

Array steering

If the propagation delays from the source of interest to each microphone are known, we may assume that the desired signal $s(p)$ reaches the microphones synchronously, i.e., that the main peaks of the acoustic channels h_m in (2.6) are synchronous. To fulfill this assumption for any source position, the propagation delays from the source to the microphones need to be compensated so that the desired source signal is time-aligned at all beamformer inputs. This operation is called array steering.

Let us denote the propagation delay from the source to the microphone x_m as τ_m. For an implementation with causal filters, one considers the positive delays $\tau'_m = -\tau_m + D', m = 1, \ldots, M$, with D' greater than $\max_m \tau_m$. The time-aligned input signals are obtained by replacing the microphone signals $x_m(p)$ with $x_m(p - \tau'_m)$:

$$x_m(p - \tau'_m) \rightarrow x_m(p). \tag{3.23}$$

In (3.23), we compensate the individual propagation delays. The delays τ_m are set according to the assumed position of the source relative to the microphones. Since the delays τ_m are a continuous quantity, the compensation delays τ'_m are generally no multiple of the sampling period and fractional-delay filters are necessary [64]. In the following we assume that the array is correctly steered to the position of the source of interest, that is, we assume that the desired source signal is time-aligned at all beamformer inputs.

A spatially constrained LCMV

For compact microphone arrays, we use the following $ML \times L$ constraint matrix and $L \times 1$ response vector $\mathbf{c}$:

$$\mathbf{C} = [\mathbf{I}_{L \times L}, \ldots, \mathbf{I}_{L \times L}]^{\mathrm{T}}, \tag{3.24}$$
$$\mathbf{c} = \boldsymbol{\delta}_D. \tag{3.25}$$

The physical meaning of this constraint is based on the assumption of a free-field acoustic propagation. Under this assumption, the impulse responses from the source of interest to the *steered* input signals are synchronous unit impulses and we may set $\boldsymbol{\delta}_{\tau_m, \theta f_s} = \boldsymbol{\delta}_0$ in (2.27). Then the spatial response (2.27) in the steered direction $\boldsymbol{\theta}_0$ may be written very simply as

$$\mathbf{g}(\mathbf{w}, \boldsymbol{\theta}_0) = \sum_{m=1}^{M} \mathbf{w}_m(p). \tag{3.26}$$

Also, observing that the constraint

$$\sum_{m=1}^{M} \mathbf{w}_m(p) = \boldsymbol{\delta}_D \tag{3.27}$$

can be written as $\mathbf{C}^{\mathrm{T}}\mathbf{w}(p) = \mathbf{c}$ for $\mathbf{C}$ and $\mathbf{c}$ set as in (3.24) and (3.25), we see that (3.24) and (3.25) define a LCMV beamformer with the spatial constraint $\mathbf{g}(\mathbf{w}(p), \boldsymbol{\theta}_0) = \boldsymbol{\delta}_D$.

Fixed beamformer and blocking matrix

The fixed beamformer is set to

$$\mathbf{w_0} \triangleq \frac{1}{M} [\boldsymbol{\delta}_D, \ldots, \boldsymbol{\delta}_D].$$

(3.28)

$\boldsymbol{\delta}_D$ denotes a D-delayed unit impulse. It may be easily verified that $\mathbf{w_0}$ in (3.28) satisfies (3.14) for the constraint in (3.24) and (3.25). As a delay-and-sum beamformer, the fixed beamformer should produce an enhanced desired signal from the microphone signals. Using (3.28), the output of the fixed beamformer in (3.17) becomes $x_0(p) = \frac{1}{M} \sum_{m=1}^{M} x_m(p - D)$. We note that more sophisticated fixed beamformers such as filter-and-sum beamformers could be used [48].

Conversely, the blocking matrix is designed to cancel the signals coming from the steered direction. The $ML \times M'L$ blocking matrix $\mathbf{B}$ may be set to

$$\mathbf{B} \triangleq \frac{1}{M} \begin{bmatrix} -\mathbf{I}_{L \times L} & \cdots & -\mathbf{I}_{L \times L} \\ (M-1)\mathbf{I}_{L \times L} & \ddots & \\ & \ddots & \ddots & \vdots \\ & & -\mathbf{I}_{L \times L} \\ \vdots & \ddots & \\ (M-1)\mathbf{I}_{L \times L} & \cdots & (M-1)\mathbf{I}_{L \times L} \end{bmatrix}.$$

(3.29)

It may be easily verified that $\mathbf{B}$ in (3.29) satisfies (3.15) for the constraint in (3.24) and (3.25). Other implementations with a zero spatial response in the steered direction are possible (for example by pairwise subtracting the time-aligned microphone signals as in the original implementation of the GSC [42]). In terms of spatial response, it is designed to have a zero spatial response in the steered direction, which is also called "null-steering beamforming." Ideally, the blocking matrix should cancel the target signal components from the input signals. Using (3.29), the blocking matrix output in (3.17) becomes

$$x_{B,m'}(p) = x_{B,m'}(p) - \frac{1}{M} \sum_{m=1}^{M} x_m(p - D),$$

(3.30)

$$= x_{B,m'}(p) - x_0(p)$$

(3.31)

for $m' = 1, \ldots, M'$.

3.3.2 Constraint for Distributed Microphone Array

In the case of the distributed microphone array, we assume that each source s_n is placed closest to its microphone x_n for $n = 1, \ldots, N$. Then the target signal reference and the interference reference signals are directly provided

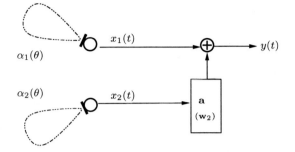

Fig. 3.2. The beamformer structure for two directional microphones is a simple adaptive interference canceler (AIC). The interference canceler coefficients are still denoted by $\mathbf{a}(p)$ for consistency with the GSC structure, but we have $\mathbf{a}(p) = \mathbf{w}_2(p)$. The mth microphone is placed closest to the mth source (for $m = 1, 2$) so that the delay D may be set to $D = 0$

by the microphones and we can define the target signal reference and the interference reference signals as follows:

$$x_0(p) \triangleq x_1(p), \tag{3.32}$$

$$x_{B,m'}(p) \triangleq x_{m'+1}(p), \quad m' = 1, \ldots, M-1. \tag{3.33}$$

In terms of fixed beamformer and blocking matrix, (3.32) and (3.33) correspond to:

$$\mathbf{w}_0 = \left[\boldsymbol{\delta}_0^{\mathrm{T}}, \mathbf{0}_{L\times1}^{\mathrm{T}} \cdots \mathbf{0}_{L\times1}^{\mathrm{T}} \right]^{\mathrm{T}} \quad \mathbf{B} = \begin{bmatrix} \mathbf{0}_{L\times(M-1)L} \\ \mathbf{I}_{(M-1)L\times(M-1)L} \end{bmatrix}. \tag{3.34}$$

In terms of constraint matrix and response vector, these settings correspond to:

$$\mathbf{C} = \left[\mathbf{I}_{L\times L} \, \mathbf{0}_{L\times L} \cdots \mathbf{0}_{L\times L} \right]^{\mathrm{T}} \quad \mathbf{c} = \boldsymbol{\delta}_0. \tag{3.35}$$

In contrast to (3.28), no delay is introduced. With the appropriate positioning of the microphones relative to the acoustic sources, causality constraints may be set on the separation system. With (3.32) and (3.33), the GSC reduces to Widrow's interference canceler, which is shown in Fig. 3.2 (see also Fig. 2.3). However, we maintain the distinction between the beamformer input signals $x_m(p)$ and the target $x_0(p)$ and interference reference signals $x_{B,m'}$ for distributed microphone arrays. This allows to keep the same notations for compact and distributed microphone arrays.

3.4 The Target Signal Cancelation Problem

We have seen in Sect. 3.2 that the LCMV optimization problem may be reformulated into an unconstrained fashion, with the Wiener solution in (3.22) as the optimal interference canceler. However, as we have seen from the gradient expression in (3.21), the Wiener solution also decorrelates the interference reference $\mathbf{x}_B(p)$ and the beamformer output $y(p)$, in addition to minimizing the output signal power. As a consequence, if the target signal is present in (or correlated with) $\mathbf{x}_B(p)$, the LCMV adaption criterion will tend to decorrelate the target signal and the beamformer output. In other words, the LCMV adaption criterion will tend to remove the target signal from the beamformer output. Practically, this means that an undesired cancelation of the target signal occurs if the adaptation of $\mathbf{a}(p)$ is carried out during target activity. This obviously undesired effect is referred to as the target signal cancelation problem. In the following, we describe the target signal cancelation problem mathematically and discuss alternative implementations of the GSC that tackle this problem.

3.4.1 The Energy-Inversion Effect

We illustrate the target signal cancelation problem for the case of two microphones ($M = 2$). The analysis is carried out in the discrete-time fourier transform (DTFT) domain. The target signal is denoted by $S_1(\omega)$ and the interferer by $S_2(\omega)$. The output of the fixed beamformer $X_0(\omega)$ consists of the filtered target signal $H_{01}(\omega)S_1(\omega)$ and the filtered interferer $H_{02}(\omega)S_2(\omega)$. The output of the blocking matrix contains the filtered interferer $H_{B2}(\omega)S_2(\omega)$ and the leakage of the desired signal $H_{B1}(\omega)S_1(\omega)$. The filters H_{0i}, H_{Bi} for $i = 1, 2$ combine acoustic transfer functions and GSC spatial preprocessing filters. $X_0(\omega)$ and $X_B(\omega)$ are accordingly defined as

$$
\begin{aligned}
X_0(\omega) &\triangleq H_{01}(\omega)S_1(\omega) + H_{02}(\omega)S_2(\omega), \\
X_B(\omega) &\triangleq H_{B1}(\omega)S_1(\omega) + H_{B2}(\omega)S_2(\omega).
\end{aligned}
\tag{3.36}
$$

The DTFT of $\mathbf{a}(p)$ is denoted by $A(\omega)$ and the beamformer output is given by

$$
Y(\omega) = X_0(\omega) + A(\omega)X_B(\omega).
\tag{3.37}
$$

We denote $\mathbf{E}\left\{|S_i(\omega)|^2\right\}$ by $\sigma_i^2(\omega)$. Assuming that $S_1(\omega)$ and $S_2(\omega)$ are uncorrelated (i.e., that $\mathbf{E}\left\{S_1(\omega)S_2^*(\omega)\right\} = 0$), the Wiener solution (3.22) may be written in the DTFT-domain as follows[2]:

$$
A^* = -e^{-i\omega D}\frac{H_{01}H_{B1}^*\sigma_1^2 + H_{02}H_{B2}^*\sigma_2^2}{|H_{B1}|^2\sigma_1^2 + |H_{B2}|^2\sigma_2^2}.
\tag{3.38}
$$

Substituting A from (3.38) and X_0, X_B from (3.36) into (3.37), the output can be written after some manipulations as

$$
Y = e^{-i\omega D}\frac{H_{01}H_{B2} - H_{02}H_{B1}}{|H_{B1}|^2\sigma_1^2 + |H_{B2}|^2\sigma_2^2}\left(S_1H_{B2}^*\sigma_2^2 - S_2H_{B1}^*\sigma_1^2\right).
\tag{3.39}
$$

---

[2] We omit the argument ω.

In this context, the signal-to-interference ratio at the blocking matrix output, denoted by SIR_{X_B}, may be defined as

$$\text{SIR}_{X_B} \triangleq \frac{\sigma_1^2 |H_{B1}|^2}{\sigma_2^2 |H_{B2}|^2}. \tag{3.40}$$

The signal-to-interference ratio at the beamformer output, SIR_{out}, can be derived directly from (3.39) as

$$\text{SIR}_{\text{out}} = \frac{\sigma_2^2 |H_{B2}|^2}{\sigma_1^2 |H_{B1}|^2}. \tag{3.41}$$

Finally, we have

$$\text{SIR}_{\text{out}} = (\text{SIR}_{X_B})^{-1}. \tag{3.42}$$

Therefore, the signal-to-interference ratio at the output is the interference-to-signal ratio at the interferer reference, i.e., the output of the blocking matrix. This phenomenon is called the energy inversion effect [28].

3.4.2 Countermeasures: On the Necessity of a Double-Talk Detector

The target signal cancelation problem is well-known [91]. In the following we discuss alternative implementations of the GSC that tackle this problem.

The straightforward and most widely used remedy consists in detecting periods of target activity, and interrupting the adaptation during these periods [27]. In the context of multiple speakers, we call the control mechanism[3] a double-talk detector (DTD). "Double-Talk" refers to the situation where the desired source and the interferer are active simultaneously. As in echo cancelation, the design of a reliable DTD has revealed to be delicate [89]. Van Compernolle proposed a method based on the power of the microphone signals, assuming that the source of interest is significantly louder than the other sources [27]. In many implementations, the input SIR estimation is based on the power ratio of the delay-and-sum fixed beamformer and blocking matrix outputs [54, 44]. This SIR estimate may be compared to a fixed decision threshold [54]. However, it is not clear how to set a universal threshold to provide an accurate double-talk detection in various nonstationary conditions. (This issue is illustrated on an example in Appendix A.)

These detection strategies lead to an all-or-nothing adaptation control, for which the value of the decision threshold may influence the performance significantly. Moreover, since the DTD should stop the adaptation when both the target speaker and the interferer are active, it reduces the tracking capabilities of the adaptive interference canceler.

---

[3] In the context of a single speaker under noisy conditions (e.g., road noise in car interior), the term voice activity detector (VAD) is also widely used.

Another complementary approach to reduce the target signal cancelation is to combat its cause. As shown by Affes et al. [2], this may be done by adapting the constraint in (3.5) to the actual acoustic channels instead of relying upon an oversimplified propagation model [2, 37, 53]. The subsequent modifications affect mainly the blocking matrix, which is responsible for target leakage. For example, the Robust GSC (RGSC) proposed by Hoshuyama et al. [53] uses an adaptive blocking matrix which cancels the desired signal components using a set of filters similar to interference cancelers. The mechanism they proposed has been analyzed and extended to the frequency domain [47]. It provides more robustness against target leakage but still requires a DTD [2, 53]. Note that the approach by Gannot et al. may perform without control mechanism, but is restricted to stationary interference (or noise) signals [37] and is not adapted to the case of interfering speech.

As a summary, we observe that the complementary countermeasures may not dispense from a DTD-based control mechanism. In the next chapters, for comparison purposes, the RGSC proposed by Hoshuyama et al. is considered [53].[4]

3.5 Summary and Conclusion

This chapter introduced fundamental concepts in LCMV beamforming. LCMV beamforming consists in adjusting the filter coefficients to minimize the output signal power while maintaining a certain linear constraint. LCMV beamformers may be formulated as GSCs which include a target signal reference and one or several interference references, and which allows for use of unconstrained optimization algorithms. Also, GSC beamformers may be easily designed for both compact and distributed microphone arrays.

Unfortunately, minimizing the output signal power leads to the target-cancelation problem, since in real environment the desired signal always leaks into the interferer references. This is due to several factors, which cannot be avoided in practice, such as reverberation, array imperfections, or steering errors. For these reasons, LCMV beamformers may cancel the target signal if the adaptation is not stopped during the periods of target activity. In the context of multiple concurrent speakers, the adaptation control involves a double-talk detector (DTD).

The design of a reliable DTD may be delicate and most control strategies are all-or-nothing triggers involving decision thresholds, whose value may influence the performance significantly. Also, the adaptation is stopped when both the target speaker and the interferer are active ("Double-Talk"). This reduces the tracking capabilities of the adaptive interference canceler, in particular for overlapping target and interferer speech. This motivates our investigations in Chap. 4 to design an adaptive beamforming algorithm without detection, where the adaption is realized continuously.

---

[4] More details are given in Appendix C.

To summarize, the important result of this chapter is the following: A classical approach in multichannel speech enhancement is LCMV beamforming implemented as a GSC. LCMV beamforming is a supervised, informed approach to the speaker separation. Extraneous information is required at two levels:

- Firstly, the position of the target source is used at the algorithmic (compact array) or at the physical (distributed array) level.
- Secondly, the periods of time where the target signal level is high relative to that of the interferer signal must be detected.

4

Implicit Adaptation Control for Beamforming

In Chap. 3, we have presented the statistically optimum LCMV beamforming and the Wiener solution (3.22). The latter involves the second-order statistics of the input signals, which are unknown in general. They might be estimated from the data under the assumption of stationary ergodic input microphone signals. However, in our application, the acoustic environment may change over time, for example when the speakers move. Furthermore, speech signals are nonstationary, requiring an adaptive approach.

On the one hand, continuous adaptation is desirable to find and to track the time-variant optimum filter coefficients. On the other hand, the filter estimation needs to be carried out only when the interferer is dominant relative to the target. This latter requirement may be satisfied with an *explicit* adaptation control based on a double-talk detector (DTD): We adapt only when some estimate of the input SIR is below a certain threshold. However, as we have seen in Chap. 3, the design of a robust and reliable DTD-based adaptation control is often difficult.

This motivates the focus on another approach where the adaptation control is realized *implicitly* and continuously. The chapter is organized as follows: Section 4.1 presents the normalized least mean square algorithm (NLMS), a widely used adaptive algorithm with a normalized step-size. Section 4.2 introduces a time-variant step-size that takes on the adaptation control. This yields an implicitly controlled LMS algorithm (ILMS). In Sect. 4.3, we examine the behavior of the ILMS algorithm theoretically. In Sect. 4.4, we set a customary constraint to further limit the target signal cancelation. In Sect. 4.5, ILMS and NLMS are compared experimentally.

4.1 Adaptive Interference Canceler

Let us briefly recall the notations. We consider the GSC shown in Fig. 3.1 as the base architecture. The input signals are:

1. A target reference signal, which is the output of the delay-and-sum fixed beamformer and is given by $x_0(p) \triangleq \mathbf{w}_0^T \mathbf{x}(p)$
2. $M - 1$ interferer references, which are the outputs of the blocking matrix $\mathbf{B}$ and are stacked in the $(M - 1)L \times 1$ vector $\mathbf{x}_B(p) \triangleq \mathbf{B}^T \mathbf{x}(p)$

The system output $y(p)$ is defined as $y(p) \triangleq x_0(p-D) + \mathbf{a}^T(p)\mathbf{x}_B(p)$, where the vector $\mathbf{a}(p)$ contains the $(M - 1)L$ coefficients of the interference canceler. The optimal time-variant interference canceler $\mathbf{a}_{\mathrm{opt}}(p)$ minimizes the interferer signal power at the output $y(p)$ while letting the target signal pass. Now the question is how to adapt the interference canceler $\mathbf{a}(p)$ to track $\mathbf{a}_{\mathrm{opt}}(p)$.

We may distinguish two categories of adaptive algorithms: (1) block-wise algorithms which estimate the filter periodically using a block of $K \geq L$ input samples according to a closed-form equation such as (3.22), and (2) sample-wise algorithms which update iteratively the filter coefficients after each new input sample such as in (2.21). In this chapter, only the latter category of sample-wise adaptive algorithms is considered.

The LMS algorithm as a gradient descent

A starting point for the adaptation of $\mathbf{a}(p)$ is the least-mean-square (LMS) algorithm with step-size μ_{LMS}. Consider the gradient descent for the cost function $J(p) \triangleq y^2(p)$ with N_{iter} iterations for each time p. This gradient descent may be written as

$$\mathbf{a}(n + 1, p) \triangleq \mathbf{a}(n, p) - \frac{\mu_{\mathrm{LMS}}}{2} \frac{\partial J(p)}{\partial \mathbf{a}}, \tag{4.1}$$

with $\mathbf{a}(0, p) \triangleq \mathbf{a}(N_{\mathrm{iter}}, p-1)$ as initialization. If we set $N_{\mathrm{iter}} = 1$, we may drop the iteration index n in (4.1):

$$\mathbf{a}(p + 1) = \mathbf{a}(p) - \frac{\mu_{\mathrm{LMS}}}{2} \frac{\partial J(p)}{\partial \mathbf{a}}, \tag{4.2}$$

and obtain the LMS algorithm [92] that updates the interference canceler coefficients for each new sample as follows:

$$\mathbf{a}(p + 1) = \mathbf{a}(p) - \mu_{\mathrm{LMS}}\, y(p)\mathbf{x}_B(p). \tag{4.3}$$

The speed of convergence, the steady-state misadjustment, and the stability are controlled by the step-size μ_{LMS} [45, 92]. Assuming wide-sense stationary signals $\mathbf{x}_B(p)$, it may be shown[1] that the mean sequence $\mathbf{E}\{\mathbf{a}(p)\}$ converges to a finite $\mathbf{a}(\infty)$ if

---

[1] A usual assumption for the LMS algorithm analysis is the so-called independence assumption. It is assumed that the elements of the vector $\mathbf{x}_B(p)$ and those of $\mathbf{x}_B(p')$ are independent if $p \neq p'$. The assumption is obviously wrong for $L > 1$, even for independent and identically distributed interference signals. However, it leads to a realistic description of the LMS behavior, and giving it up would render the analysis considerably more complex [92].

$$0 < \mu_{\text{LMS}} < \frac{2}{L \sum_{m=1}^{M-1} \mathbf{E}\left\{x_{B,m}^2(p)\right\}}. \tag{4.4}$$

The speed of convergence of the LMS algorithm depends not only on the step-size μ_{LMS} but also on the eigenvalue spread of the correlation matrix $\mathbf{R}_{\mathbf{x}_B\mathbf{x}_B}(p) = \mathbf{E}\left\{\mathbf{x}_B(p)\mathbf{x}_B^{\mathrm{T}}(p)\right\}$. Let us denote the largest and smallest eigenvalues of $\mathbf{R}_{\mathbf{x}_B\mathbf{x}_B}(p)$ by $\lambda_{\max}$ and $\lambda_{\min}$, respectively. The smaller the ratio $\lambda_{\max}/\lambda_{\min}$ is, the faster the convergence. The eigenvalue spread $\lambda_{\max}/\lambda_{\min}$ is minimal and equals one for white and uncorrelated input signals $x_{B,m}(p)$. Fast convergence may also be achieved for colored signals with, for example, the recursive least-square (RLS) algorithm [45].

Since the power of speech signals is highly time-variant, a fixed step-size μ_{LMS} will usually not stay close to the desirable upper bound in (4.4). This problem is addressed by the normalized LMS (NLMS) using a normalized step-size.

Normalized step-size

The upper bound on μ_{LMS} in (4.4) is inversely proportional to the power of the interferer reference signals $\sum_{m=1}^{M-1} \mathbf{E}\left\{x_{B,m}^2(p)\right\}$. Let us consider the normalized step-size μ_{NLMS} given by

$$\mu_{\text{NLMS}} \triangleq \mu_{\text{LMS}} L \sum_{m=1}^{M-1} \mathbf{E}\left\{x_{B,m}^2(p)\right\}. \tag{4.5}$$

According to (4.4), convergence of $\mathbf{E}\left\{\mathbf{a}(p)\right\}$ is guaranteed if the following condition is satisfied:

$$0 < \mu_{\text{NLMS}} < 2. \tag{4.6}$$

Let us estimate the input power $L \sum_{m=1}^{M-1} \mathbf{E}\left\{x_{B,m}^2(p)\right\}$ with the instantaneous estimate $\|\mathbf{x}_B(p)\|^2$, where $\|\mathbf{x}\|^2 = \mathbf{x}^{\mathrm{T}}\mathbf{x}$. From (4.3) and (4.5), we obtain the normalized LMS algorithm (NLMS) [45]:

$$\mathbf{a}(p+1) = \mathbf{a}(p) - \mu_{\text{NLMS}} \frac{y(p)\mathbf{x}_B(p)}{\|\mathbf{x}_B(p)\|^2}. \tag{4.7}$$

Since some leakage of the target signal in the interferer reference $\mathbf{x}_B$ always exists, the interference canceler $\mathbf{a}(p)$ converges to the optimal $\mathbf{a}_{\text{opt}}(p)$ only if the target signal is zero. Otherwise, the adaptation should be slowed down with a time-varying, smaller step-size or stopped with $\mu_{\text{NLMS}} = 0$, depending on the input SIR. In the next section, we attempt to design such a time-varying step-size.

4.2 Implicit Adaptation Control

Heuristic introduction

The normalization term of the NLMS update in (4.7) is given by $1/\|\mathbf{x}_B(p)\|^2$. This normalization assures the stability of the algorithm, but does not lead to a faster adaptation in favorable conditions, namely, when the input SIR is low. Another shortcoming of the NLMS algorithm is that the steady-state misadjustment increases with the target signal power, even if the target signal does not leak into the interferer reference [41]. Therefore, it makes sense to use a large step-size when the target signal power is low. To this end, let us define the output vector $\mathbf{y}(p) \triangleq (y(p), \ldots, y(p - L + 1))^{\mathrm{T}}$ and consider $\|\mathbf{y}(p)\|^2/L$ as an estimate of the target signal power. When the target signal power is low, a large adaptation term is obtained by replacing $1/\|\mathbf{x}_B(p)\|^2$ in (4.7) with $1/\|\mathbf{y}(p)\|^2$. This also yields a small adaptation term when the target signal power is high, which reduces the risk of target signal cancelation. In other words, an *implicitly* controlled adaptation is obtained by replacing the NLMS algorithm with

$$\mathbf{a}(p + 1) = \mathbf{a}(p) - \mu_0 \frac{y(p)\mathbf{x}_B(p)}{(M - 1)\|\mathbf{y}(p)\|^2}. \tag{4.8}$$

The algorithm (4.8) includes an implicit adaptation control, and in the following we refer to it as Implicit LMS (ILMS) with step-size μ_0. However, ILMS as given in (4.8) is not stable, since $\|\mathbf{y}(p)\|^2$ might become very small.

Stability conditions

Unfortunately, the condition $0 < \mu_0 < 2$ does not guarantee the stability of (4.8) in the mean. A common approach to ensure stability is to increase the denominator of the update term in (4.8) by a fixed regularization term $\delta > 0$ [44], as follows:

$$\mathbf{a}(p + 1) = \mathbf{a}(p) - \mu_0 \frac{y(p)\mathbf{x}_B(p)}{\|\mathbf{y}(p)\|^2 + \delta}. \tag{4.9}$$

However, this fixed regularization scheme generally reduces the convergence speed.

In the following, we propose an alternative approach. The ILMS algorithm in (4.8) may be seen as a special version of the NLMS algorithm (4.7) with a time-varying step-size. Replacing the step-size μ_{NLMS} by

$$\mu_0 \frac{\|\mathbf{x}_B(p)\|^2}{(M - 1)\|\mathbf{y}(p)\|^2} \tag{4.10}$$

in (4.7), we directly obtain the ILMS equation (4.8). Hence we can consider the domain of stability given in (4.6). Using the variable step-size in (4.10), the domain of stability (4.6) may be written as

$$0 < \mu_0 \frac{\|\mathbf{x}_B(p)\|^2}{(M-1)\|\mathbf{y}(p)\|^2} < 2. \tag{4.11}$$

If the condition (4.11) is not satisfied, we may simply perform the adaptation with the standard NLMS and the step-size μ_0. In practice we may consider a stability condition that is more conservative than (4.11) with a maximal step-size $\mu_{\max} < 2$ and

$$0 < \mu_0 \frac{\|\mathbf{x}_B\|^2}{(M-1)\|\mathbf{y}\|^2} < \mu_{\max}. \tag{4.12}$$

To summarize, the ILMS algorithm can be written as

$$\mathbf{a}(p+1) = \mathbf{a}(p) - \begin{cases} \mu_0 \frac{y(p)\mathbf{x_B}(p)}{(M-1)\|\mathbf{y}(p)\|^2} & \text{if } \mu_0 \frac{\|\mathbf{x}_B(p)\|^2}{(M-1)\|\mathbf{y}(p)\|^2} < \mu_{\max}, \\ \mu_0 \frac{y(p)\mathbf{x_B}(p)}{\|\mathbf{x}_B(p)\|^2} & \text{otherwise.} \end{cases} \tag{4.13}$$

Note that the ILMS and NLMS algorithms yield the same computational complexity.

4.3 Analysis of the ILMS Algorithm

Having introduced the ILMS algorithm heuristically in Sect. 4.2, we attempt in this section to motivate this algorithm with theoretical arguments. In Sect. 4.3.1 we will show that the ILMS algorithm is an approximation of the NLMS algorithm with a time-variant, optimal step-size, under the assumption that the target signal does not leak into the interference reference. In Sect. 4.3.2 we relax this assumption and provide an analysis of the mean trajectory of the ILMS algorithm for a simplified source model.

4.3.1 Linking ILMS to the NLMS with Optimal Step-Size

Interference canceler mismatch

A useful preliminary is to introduce the mismatch between the actual and the optimal interference canceler, as in system identification. The mismatch at time p is defined as

$$\mathbf{m}(p) \triangleq \mathbf{a}(p) - \mathbf{a}_{\text{opt}}(p). \tag{4.14}$$

We denote the contribution of the desired signal in $\mathbf{x}(p)$ by $\mathbf{d}(p)$ and that of the interference by $\mathbf{n}(p)$, that is,

$$\mathbf{x}(p) = \mathbf{d}(p) + \mathbf{n}(p). \tag{4.15}$$

We now present an expression of the output $y(p)$ as a function of the mismatch $\mathbf{m}(p)$:

$$y(p) = \mathbf{w}_0^{\mathrm{T}}\mathbf{x}(p - D) + \left(\mathbf{m}^{\mathrm{T}}(p) + \mathbf{a}_{\mathrm{opt}}^{\mathrm{T}}(p)\right)\mathbf{B}^{\mathrm{T}}\mathbf{x}(p), \tag{4.16}$$

$$= \mathbf{w}_0^{\mathrm{T}}\left(\mathbf{d}(p - D) + \mathbf{n}(p - D)\right)$$
$$+ \left(\mathbf{m}^{\mathrm{T}}(p) + \mathbf{a}_{\mathrm{opt}}^{\mathrm{T}}(p)\right)\mathbf{B}^{\mathrm{T}}\left(\mathbf{d}(p) + \mathbf{n}(p)\right). \tag{4.17}$$

It is then assumed that $\mathbf{a}_{\mathrm{opt}}(p)$ perfectly cancels the interferer at the output, i.e., that

$$\mathbf{w}_0^{\mathrm{T}}\mathbf{n}(p - D) + \mathbf{a}_{\mathrm{opt}}^{\mathrm{T}}(p)\mathbf{B}^{\mathrm{T}}\mathbf{n}(p) = 0.$$

This assumption is not critical if we are considering the transient behavior and not the steady state of the adaptation. Combining this assumption with (4.17) yields

$$y(p) = \mathbf{w}_0^{\mathrm{T}}\mathbf{d}(p - D) + \mathbf{a}_{\mathrm{opt}}^{\mathrm{T}}(p)\mathbf{B}^{\mathrm{T}}\mathbf{d}(p) + \mathbf{m}^{\mathrm{T}}(p)\mathbf{B}^{\mathrm{T}}(\mathbf{n}(p) + \mathbf{d}(p)). \tag{4.18}$$

Substituting $\mathbf{n}(p) = \mathbf{x}(p) - \mathbf{d}(p)$ in (4.18), we obtain

$$y(p) = \mathbf{w}_0^{\mathrm{T}}\mathbf{d}(p - D) + \mathbf{a}_{\mathrm{opt}}^{\mathrm{T}}(p)\mathbf{B}^{\mathrm{T}}\mathbf{d}(p) + \mathbf{m}^{\mathrm{T}}(p)\mathbf{B}^{\mathrm{T}}\mathbf{x}(p). \tag{4.19}$$

We define $b(p)$ as the target signal at the output when $\mathbf{a}(p) = \mathbf{a}_{\mathrm{opt}}(p)$, i.e., when the interferer is canceled.

$$b(p) \triangleq \mathbf{w}_0^{\mathrm{T}}\mathbf{d}(p - D) + \mathbf{a}_{\mathrm{opt}}^{\mathrm{T}}(p)\mathbf{B}^{\mathrm{T}}\mathbf{d}(p). \tag{4.20}$$

With this definition of $b(p)$, (4.19) can be rewritten as

$$y(p) = b(p) + \mathbf{m}^{\mathrm{T}}(p)\mathbf{x}_B(p). \tag{4.21}$$

Let us assume that the optimal interference canceler varies slowly, i.e., that $\mathbf{a}_{\mathrm{opt}}(p+1) = \mathbf{a}_{\mathrm{opt}}(p)$. Then, the NLMS adaptation (4.7) can be written using the mismatch $\mathbf{m}(p)$ and the step-size μ_{NLMS} as

$$\mathbf{m}(p + 1) = \mathbf{m}(p) - \mu_{\mathrm{NLMS}}\frac{y(p)\mathbf{x}_B(p)}{\|\mathbf{x}_B(p)\|^2}, \tag{4.22}$$

$$= \mathbf{m}(p) - \mu_{\mathrm{NLMS}}\frac{b(p)\mathbf{x}_B(p) + \mathbf{x}_B(p)\mathbf{x}_B^{\mathrm{T}}(p)\mathbf{m}(p)}{\|\mathbf{x}_B(p)\|^2}. \tag{4.23}$$

Optimal step-size derivation

We now derive a time-variant optimal step-size $\mu(p)$. The optimality criterion $J(\mu(p))$ that we consider for this derivation is the expected quadratic norm of the mismatch at time $p + 1$:

$$J(\mu(p)) \triangleq \mathbf{E}\left\{\|\mathbf{m}(p + 1)\|^2\right\}. \tag{4.24}$$

Denoting the time-variant step-size by $\mu(p)$ instead of μ_{NLMS} in (4.22) and substituting $\mathbf{m}(p+1)$ from (4.22) into (4.24) yields[2]

$$J(\mu) = \mathbf{E}\left\{\|\mathbf{m}\|^2\right\} + \mu^2 \mathbf{E}\left\{\frac{y^2}{\|\mathbf{x}_B\|^2}\right\} - 2\mu \mathbf{E}\left\{\frac{y\mathbf{m}^{\mathrm{T}}\mathbf{x}_B}{\|\mathbf{x}_B\|^2}\right\}. \qquad (4.25)$$

The derivative of J with respect to $\mu(p)$ is

$$\frac{\partial J(\mu)}{\partial \mu} = 2\left(\mu \mathbf{E}\left\{\frac{y^2}{\|\mathbf{x}_B\|^2}\right\} - \mathbf{E}\left\{\frac{y\mathbf{m}^{\mathrm{T}}\mathbf{x}_B}{\|\mathbf{x}_B\|^2}\right\}\right). \qquad (4.26)$$

Since the cost function $J(\mu)$ is quadratic in μ, it has only one minimum. Solving for $\frac{\partial J(\mu(p))}{\partial \mu(p)} = 0$ yields a closed formula for this minimum. The optimal step-size $\mu_{\mathrm{opt}}(p)$ is given by

$$\mu_{\mathrm{opt}}(p) = \frac{\mathbf{E}\left\{\frac{y(p)\mathbf{m}^{\mathrm{T}}(p)\mathbf{x}_B(p)}{\|\mathbf{x}_B(p)\|^2}\right\}}{\mathbf{E}\left\{\frac{y^2(p)}{\|\mathbf{x}_B(p)\|^2}\right\}}. \qquad (4.27)$$

Unfortunately, this optimal step-size cannot be computed in practice, since the mismatch $\mathbf{m}(p)$ and the expectation $\mathbf{E}\{\}$ are unknown. Approximations are necessary for a practical implementation, leading to a *pseudo*optimal step-size.

Approximating the optimal step-size

First, we assume that $\|\mathbf{x}_B(p)\|^2$ may be approximated by a deterministic variable, as in [68]. Then, the term $\|\mathbf{x}_B(p)\|^2$ may be factored out of (4.27). Therefore, a first approximation of the optimal step-size is

$$\mu_{\mathrm{opt}}(p) \approx \frac{\mathbf{E}\left\{y(p)\mathbf{m}^{\mathrm{T}}(p)\mathbf{x}_B(p)\right\}}{\mathbf{E}\left\{y^2(p)\right\}}. \qquad (4.28)$$

Substituting $y(p)$ from (4.21) into (4.28) yields

$$\mu_{\mathrm{opt}}(p) \approx \frac{\mathbf{E}\left\{b(p)\mathbf{m}^{\mathrm{T}}(p)\mathbf{x}_B(p)\right\} + \mathbf{E}\left\{|\mathbf{m}^{\mathrm{T}}(p)\mathbf{x}_B(p)|^2\right\}}{\mathbf{E}\left\{y^2(p)\right\}}. \qquad (4.29)$$

The signal $b(p)$ depends on the target signal and on the optimal filter coefficients, which are unknown. The correlation term $\mathbf{E}\left\{b(p)\mathbf{m}^{\mathrm{T}}(p)\mathbf{x}_B(p)\right\}$ seems difficult to estimate as it depends on the quantities $b(p)$ and $\mathbf{m}(p)$, which are unknown. For this reason, we need to assume that the target source is silent or does not leak into the interference reference, which implies $b(p) = 0$ and $\mathbf{E}\left\{b(p)\mathbf{m}^{\mathrm{T}}(p)\mathbf{x}_B(p)\right\} = 0$.

---

[2] The time index p is omitted for the sake of readability.

Assuming a white signal $\mathbf{x}_B(p)$, the term $\mathbf{E}\left\{|\mathbf{m}^{\mathrm{T}}(p)\mathbf{x}_B(p)|^2\right\}$ in (4.29) may be approximated by

$$\mathbf{E}\left\{|\mathbf{m}^{\mathrm{T}}(p)\mathbf{x}_B(p)|^2\right\} \approx \mathbf{E}\left\{\|\mathbf{m}(p)\|^2\right\} \frac{\|\mathbf{x}_B(p)\|^2}{(M-1)L}. \tag{4.30}$$

Note that the assumption of a white signal $\mathbf{x}_B(p)$ seems rather unrealistic, since speech signals and road noise are typically stronger for low frequencies.

In system identification problems like echo cancelation, elaborate techniques may provide an estimation of $\mathbf{E}\left\{\|\mathbf{m}(p)\|^2\right\}$ [44]. Some may also be applicable in beamforming, for example if certain elements of $\mathbf{a}_{\mathrm{opt}}(p)$ are *a priori* known to be zero. However, we found it much simpler and very efficient to approximate $\mathbf{E}\left\{\|\mathbf{m}(p)\|^2\right\}$ with a constant μ_0:

$$\mathbf{E}\left\{\|\mathbf{m}(p)\|^2\right\} \approx \mu_0. \tag{4.31}$$

Combining (4.30) with (4.31) yields

$$\mathbf{E}\left\{|\mathbf{m}^{\mathrm{T}}(p)\mathbf{x}_B(p)|^2\right\} \approx \mu_0 \frac{\|\mathbf{x}_B(p)\|^2}{(M-1)L}. \tag{4.32}$$

Finally, the output variance $\sigma_y^2(p) = \mathbf{E}\left\{y^2(p)\right\}$ is estimated by $\|\mathbf{y}(p)\|^2/L$. These approximations yield the pseudooptimal step-size

$$\mu(p) = \mu_0 \frac{\|\mathbf{x}_B(p)\|^2}{(M-1)\|\mathbf{y}(p)\|^2}. \tag{4.33}$$

Substituting $\mu(p)$ from (4.33) into the NLMS equation (4.7), we obtain the ILMS equation (4.8). Hence, the ILMS algorithm introduced heuristically in (4.8) has been linked to the NLMS with a pseudooptimal step-size.

4.3.2 ILMS Transient Behavior and Stability

In the previous section we had to assume that the target signal does not leak into the interference reference. In this section we relax this assumption and provide an analysis of the mean trajectory of the ILMS algorithm for a simplified source model which accounts for target leakage into the interference reference. Additionally, this section provides an interpretation of the ILMS stability condition (4.11).

Signal model

Let us first introduce a simple signal model and the notations on which the further derivations are based. The sample index p may be omitted for notational convenience. The variance of the target at the output is denoted by

σ_1^2 and the variance of the interferer at $\mathbf{x}_B$ is denoted by σ_2^2. The variance of $\mathbf{x}_B$ and y are composed of target and interferer components as shown in (4.34) and (4.35):

$$\mathbf{E}\left\{\|\mathbf{x}_B\|^2\right\} = (M-1)L\left(\varepsilon_{\text{leakage}}\sigma_1^2 + \sigma_2^2\right), \tag{4.34}$$

$$\mathbf{E}\left\{\|\mathbf{y}\|^2\right\} = L\sigma_y^2 = L\left(\sigma_1^2 + \varepsilon_{\text{mismatch}}\sigma_2^2\right). \tag{4.35}$$

The factor $\varepsilon_{\text{leakage}} \geq 0$ represents the amount of target leakage into the interference reference. The factor $\varepsilon_{\text{mismatch}} \geq 0$ controls the interferer signal power at the beamformer output, and is roughly proportional to the mismatch $\|\mathbf{m}(p)\|^2$, according to (4.21). In (4.34) and (4.35), it is assumed that the interferer and target signals are uncorrelated. In contrast to $\varepsilon_{\text{leakage}}$, $\varepsilon_{\text{mismatch}}$ is actually a time-variant quantity and may be denoted by $\varepsilon_{\text{mismatch}}(p)$. Combining (4.34) and (4.35) with the definition (4.33) yields

$$\mathbf{E}\left\{\mu\right\} = \mu_0 \frac{\varepsilon_{\text{leakage}}\sigma_1^2 + \sigma_2^2}{\sigma_1^2 + \varepsilon_{\text{mismatch}}\sigma_2^2}. \tag{4.36}$$

Transient Convergence and Divergence

The ILMS algorithm is similar to the NLMS algorithm in the sense that they share the same gradient direction, $y(p)\mathbf{x}_B(p)$. Both algorithms converge (resp. diverge) during target silence (resp. activity). However, their speed of convergence and divergence differ. The goal in this section is to obtain a quantitative estimation of the transient convergence speed of ILMS relative to NLMS. To this end, we will study the evolution of expected mismatch $\mathbf{E}\left\{\mathbf{m}(p)\right\}$ at time p and derive an upper bound on the convergence speed for ILMS and NLMS, which is obtained when the target signal is zero. The divergence speed is obtained in the worst-case scenario when the interference is silent.

We make the hypothesis that the interferer reference signals are white and uncorrelated so that $\mathbf{R}_{\mathbf{x}_B\mathbf{x}_B}(p) \propto \mathbf{I}$. This approximation is uncritical for the following derivations if we consider that the target and the interference have similar spectra during their respective periods of activity. (It is possible to remove this hypothesis by diagonalizing $\mathbf{R}_{\mathbf{x}_B\mathbf{x}_B}(p)$). It is also assumed that the algorithm is stable in the mean, i.e., $\mathbf{E}\left\{\mu(p)\right\} < 2$, since otherwise, ILMS is switched to NLMS and both algorithms behave identically. Substituting $y(p)$ from (4.21) into (4.13) and taking the expectation of both sides in (4.13), we obtain

$$\mathbf{E}\left\{\mathbf{m}(p+1)\right\} = \mathbf{E}\left\{\mathbf{m}(p)\right\} - \mathbf{E}\left\{\mu(p)\right\} \frac{\mathbf{E}\left\{b(p)\mathbf{x}_B(p)\right\} + \mathbf{R}_{\mathbf{x}_B\mathbf{x}_B}\mathbf{E}\left\{\mathbf{m}(p)\right\}}{\mathbf{E}\left\{\|\mathbf{x}_B(p)\|^2\right\}}. \tag{4.37}$$

ILMS convergence

The mismatch $\mathbf{m}(p)$ converges to zero when the target signal is zero, i.e., when $\sigma_1^2 = 0$ and $\sigma_2^2 > 0$. In this case, we have $b(p) = 0$ according to (4.20). Then, the ILMS equation (4.37) becomes

$$\mathbf{E}\left\{\mathbf{m}(p+1)\right\} = \mathbf{E}\left\{\mathbf{m}(p)\right\} - \mathbf{E}\left\{\mu(p)\right\} \frac{\mathbf{R}_{\mathbf{x}_B \mathbf{x}_B} \mathbf{E}\left\{\mathbf{m}(p)\right\}}{\mathbf{E}\left\{\|\mathbf{x}_B\|^2\right\}}. \tag{4.38}$$

The mean of the pseudooptimal step-size (4.36) is $\mathbf{E}\left\{\mu(p)\right\} = \mu_0/\varepsilon_{\text{mismatch}}$ and the correlation matrix is $\mathbf{R}_{\mathbf{x}_B \mathbf{x}_B} = \sigma_2^2 \mathbf{I}$. Using (4.34), this yields

$$\mathbf{E}\left\{\mathbf{m}(p+1)\right\} = \underbrace{\left(1 - \frac{\mu_0}{(M-1)L\,\varepsilon_{\text{mismatch}}(p)}\right)}_{\text{contraction factor } \alpha(p)} \mathbf{E}\left\{\mathbf{m}(p)\right\}. \tag{4.39}$$

Let us denote the step-size normalized to the filter length and the number of interferer references by $\tilde{\mu}_0$, that is, $\tilde{\mu}_0 \triangleq \mu_0/((M-1)L)$. The contraction factor $\alpha(p) \triangleq 1 - \frac{\tilde{\mu}_0}{\varepsilon_{\text{mismatch}}(p)}$ controls the convergence of $\mathbf{E}\left\{\mathbf{m}(p)\right\}$ to zero. $\alpha(p)$ should be as close to zero as possible and its absolute value should always be smaller than one. Observe that if an estimate of $\varepsilon_{\text{mismatch}}(p)$ were available, setting a time-variant $\tilde{\mu}_0(p) = \varepsilon_{\text{mismatch}}(p)$ would lead to the fastest convergence, as could be expected from (4.31). (Under these assumptions, convergence in the mean would be achieved in one step.)

If $\alpha(p)$ were constant, i.e., if $\alpha(p) = \alpha$, then one could characterize the transient behavior by an exponential decay with time constant

$$\tau = \frac{-1}{\ln \alpha}. \tag{4.40}$$

However, as can be seen in (4.39), $\alpha(p)$ depends on the mismatch. This has the following effect: As the system adapts, $\varepsilon_{\text{mismatch}}(p)$ decreases, thus decreasing $\alpha(p)$ and increasing the rate of convergence, until $\alpha(p)$ crosses zero. If $\alpha(p)$ crossed the value -1, (4.39) would become unstable. This is prevented by explicitly checking the stability conditions (4.12). For these reasons, the transient behavior of ILMS is not very well described by a constant contraction factor, in contrast to the traditional LMS [92]. Nevertheless, it is useful to consider an upper bound of $\alpha(p)$. Assuming that $\varepsilon_{\text{mismatch}}(p) < 1$, the contraction factor is upper-bounded by

$$\alpha(p) < a_{\text{ILMS}}(\mu_0), \tag{4.41}$$

where $a_{\text{ILMS}}(\mu_0)$ is defined as a function of μ_0 by

$$a_{\text{ILMS}}(\mu_0) \triangleq 1 - \mu_0/(M-1)L. \tag{4.42}$$

In the following, we refer to $a_{\text{ILMS}}(\mu_0)$ as the ILMS convergence contraction factor. We note that $\mathbf{E}\left\{\mathbf{m}(p)\right\}$ converges to zero faster than $a_{\text{ILMS}}^p(\mu_0)$ for $p \geq 0$.

ILMS divergence

In the worst-case situation, the target is active and the interferer is silent, i.e., $\sigma_1^2 > 0$ and $\sigma_2^2 = 0$. The interferer correlation matrix $\mathbf{R}_{\mathbf{x}_B \mathbf{x}_B}$ is now approximated by $\mathbf{R}_{\mathbf{x}_B \mathbf{x}_B} = \varepsilon_{\text{leakage}} \sigma_1^2 \mathbf{I}$. In this situation, the mean mismatch $\mathbf{E}\{\mathbf{m}(p)\}$ does not converge to zero but to another point, which we denote by $\mathbf{m}_d(p)$. The update term in (4.37) vanishes at $\mathbf{E}\{\mathbf{m}(p)\} = \mathbf{m}_d(p)$, which yields

$$\mathbf{E}\{b(p)\mathbf{x}_B(p)\} + \varepsilon_{\text{leakage}} \sigma_1^2 \mathbf{m}_d(p) = 0, \tag{4.43}$$

$$\mathbf{m}_d(p) = -\frac{\mathbf{E}\{b(p)\mathbf{x}_B(p)\}}{\varepsilon_{\text{leakage}} \sigma_1^2}. \tag{4.44}$$

The mean pseudooptimal step-size (4.36) is $\mathbf{E}\{\mu(p)\} = \mu_0 \varepsilon_{\text{leakage}}$. Let us define $\mathbf{m}'(p) \triangleq \mathbf{m}(p) - \mathbf{m}_d(p)$. Rearranging (4.37) and assuming $\mathbf{m}_d(p+1) = \mathbf{m}_d(p)$, the ILMS adaptation may be written in terms of $\mathbf{m}'(p)$ using (4.34) and (4.43) as

$$\mathbf{E}\{\mathbf{m}'(p+1)\} = \mathbf{E}\{\mathbf{m}'(p)\} \left(1 - \mu_0 \frac{\varepsilon_{\text{leakage}}}{(M-1)L}\right). \tag{4.45}$$

We define the ILMS divergence contraction factor according to (4.45) as

$$b_{\text{ILMS}}(\mu_0) \triangleq 1 - \left(\mu_0/(M-1)L\right) \varepsilon_{\text{leakage}}. \tag{4.46}$$

NLMS convergence and divergence

According to (4.37), the convergence of the NLMS with constant step-size μ_{NLMS} is described in the best-case scenario ($\sigma_1^2 = 0$, $\sigma_2^2 > 0$) by

$$\mathbf{E}\{\mathbf{m}(p+1)\} = \mathbf{E}\{\mathbf{m}(p)\} (1 - \mu_{\text{NLMS}}/(M-1)L). \tag{4.47}$$

We define the NLMS convergence contraction factor accordingly as

$$a_{\text{NLMS}}(\mu_{\text{NLMS}}) \triangleq 1 - \frac{\mu_{\text{NLMS}}}{(M-1)L}.$$

In the worst-case scenario ($\sigma_1^2 > 0$, $\sigma_2^2 = 0$), the NLMS adaptation diverges to $\mathbf{m}_d(p)$. Assuming $\mathbf{m}_d(p+1) = \mathbf{m}_d(p)$ and with $\mathbf{m}'(p) = \mathbf{m}(p) - \mathbf{m}_d(p)$ the NLMS adaptation step at time p is:

$$\mathbf{E}\{\mathbf{m}'(p+1)\} = \mathbf{E}\{\mathbf{m}'(p)\} (1 - \mu_{\text{NLMS}}/(M-1)L). \tag{4.48}$$

We define the NLMS divergence contraction factor accordingly as

$$b_{\text{NLMS}}(\mu_{\text{NLMS}}) \triangleq 1 - \frac{\mu_{\text{NLMS}}}{(M-1)L}.$$

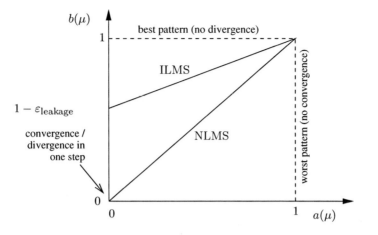

Fig. 4.1. NLMS and ILMS convergence/divergence patterns for $0 < \varepsilon_{\text{leakage}} < 1$

Comparison

To compare the joint convergence and divergence of both NLMS and ILMS algorithms, we may consider the divergence contraction factor $b(\mu)$ as a function of the convergence contraction factor $a(\mu)$, thus obtaining a convergence/divergence pattern. The best performance in terms of convergence speed would be obtained with $b(\mu) = 1$ for all $a(\mu)$: The mismatch would not be increased, even in the worst-case scenario. Conversely, the worst performance would be obtained with $a(\mu) = 1$ for all $b(\mu)$: The mismatch would not be reduced, even in the best-case scenario. The convergence/divergence patterns for the ILMS and NLMS algorithms are shown in Fig. 4.1. For the NLMS, the convergence and divergence contraction terms are equal, $b_{\text{NLMS}}(\mu) = a_{\text{NLMS}}(\mu)$, as can be seen from (4.47) and (4.48). In other words, the NLMS converges and diverges with the same speed. On the other hand, we have

$$b_{\text{ILMS}}(\mu) = a_{\text{ILMS}}(\mu)\varepsilon_{\text{leakage}} + (1 - \varepsilon_{\text{leakage}}) \qquad (4.49)$$

according to (4.42) and (4.46). If $\varepsilon_{\text{leakage}} < 1$, it may be seen in Fig. 4.1 that the convergence/divergence pattern of the ILMS is "better" than the NLMS pattern. This analysis is based on a simplistic model, e.g., only the best- and worst-case situations have been considered. Nonetheless, it suggests the good behavior of the proposed ILMS algorithm relative to NLMS.

About the Stability

In this section, we examine more closely the circumstances under which instability may be detected. Let us assume that the stability condition (4.11) is *not* satisfied in the mean. Using (4.34) and (4.35), we have

$$\mu_0 \frac{\varepsilon_{\text{leakage}}\sigma_1^2 + \sigma_2^2}{\sigma_1^2 + \varepsilon_{\text{mismatch}}\sigma_2^2} > 2. \tag{4.50}$$

Rearranging (4.50) yields

$$\varepsilon_{\text{mismatch}} < \frac{\mu_0}{2} - \frac{\sigma_1^2}{\sigma_2^2}\left(1 - \varepsilon_{\text{leakage}}\frac{\mu_0}{2}\right). \tag{4.51}$$

For sufficiently small μ_0, the term $1-\varepsilon_{\text{leakage}}\mu_0/2$ is positive and (4.51) implies

$$\begin{cases} (i) \ \varepsilon_{\text{mismatch}} < \frac{\mu_0}{2} \\ (ii) \ \frac{\sigma_1^2}{\sigma_2^2} < \frac{\frac{\mu_0}{2} - \varepsilon_{\text{mismatch}}}{1 - \varepsilon_{\text{leakage}}\frac{\mu_0}{2}} < \left(\frac{2}{\mu_0} - \varepsilon_{\text{leakage}}\right)^{-1} \end{cases} \tag{4.52}$$

Equation (4.52) tells us that the ILMS algorithm is switched to the standard NLMS only

- If the mismatch factor is smaller[3] than $\mu_0/2$ (condition (i))
- If $\frac{\sigma_1^2}{\sigma_2^2}$ is close to zero (condition (ii)), that is, if the target signal power σ_1^2 at the output is very small

Assuming that the cancelation of the target signal is not too severe (as shown in the experimental Sect. 4.5), this reveals that the stability condition (4.11) acts as a target silence detector. Therefore, the ILMS property of implicit silence detection should not be lost by switching to the standard NLMS when instability is detected.

4.4 Robustness Improvement

In most applications where prior information about the target position is available, and even if the steered DOA does not perfectly match the actual target position, the SIR is positive at the target reference signal $x_0(p)$ and negative at the interferer reference $\mathbf{x}_B(p)$. Therefore, cancelation of the target signal at the beamformer output is accompanied by a growth of the interference canceler coefficients. Based on this observation, we may replace the LMS cost function $J = \mathbf{E}\left\{y^2(p)\right\}$ by

$$J_\lambda = \mathbf{E}\left\{y^2(p)\right\} + \lambda\|\mathbf{a}\|^2, \tag{4.53}$$

for a positive weight parameter $\lambda > 0$ [45]. This penalizes large values of $\|\mathbf{a}(p)\|^2$ and partially prevents target signal cancelation. Similar to the Wiener solution (3.22), the minimum of J_λ can be obtained in closed form by setting its gradient to zero, which yields

$$\mathbf{a}(p) = -\left(\mathbf{R}_{\mathbf{x}_B\mathbf{x}_B} + \lambda\mathbf{I}\right)^{-1}\mathbf{E}\left\{x_0(p-D)\mathbf{x}_B(p)\right\}. \tag{4.54}$$

---

[3] As we will see in the experimental section, typical step-sizes are on the order of magnitude $\mu_0 = 0.01$ or smaller.

In (4.54), diagonal loading is applied on the correlation matrix $\mathbf{R}_{\mathbf{x}_B \mathbf{x}_B}$ [30]. The gradient descent for the cost function J_λ leads to the leaky LMS algorithm [45]:

$$\mathbf{a}(p+1) = (1 - \mu\lambda)\,\mathbf{a}(p) - \mu y(p)\mathbf{x}_B(p). \tag{4.55}$$

At each adaptation step, the filter coefficients are scaled down by a factor $(1 - \mu\lambda)$ that is slightly smaller than 1. The leaky LMS algorithm may also be implemented by adding random white noise to the input signals of the interference canceler [45]. Since $\mathbf{B}^T \mathbf{w}_0 = \mathbf{0}$, the white-noise gain of[4] $\mathbf{w} = \mathbf{w}_0 + \mathbf{Ba}$ is given by $\|\mathbf{w}\|^2 = \|\mathbf{w}_0\|^2 + \mathbf{a}^T \mathbf{B}^T \mathbf{Ba}$. For a blocking matrix $\mathbf{B}$ such that $\mathbf{B}^T \mathbf{B} = \mathbf{I}$, the white-noise gain becomes

$$\|\mathbf{w}\|^2 = \|\mathbf{w}_0\|^2 + \|\mathbf{a}\|^2. \tag{4.56}$$

Thus, if $\mathbf{B}^T \mathbf{B} = \mathbf{I}$ then J_λ simultaneously minimizes the output signal power and the white-noise gain of $\mathbf{w}$.

A drawback of this approach is that J_λ for $\lambda > 0$ yields a slower initial convergence than J_0. An alternative consists in restraining $\|\mathbf{a}(p)\|^2$ with the quadratic inequality constraint

$$\|\mathbf{a}(p)\|^2 < a_{\mathrm{QIC}}^2, \tag{4.57}$$

where a_{QIC} is a positive constant [30]. The constraint in (4.57) may be efficiently implemented as a projection on a ball of radius a_{QIC} [55]:

$$\mathbf{a}(p) \leftarrow \frac{a_{\mathrm{QIC}}}{\|\mathbf{a}(p)\|}\mathbf{a}(p) \quad \text{if } \|\mathbf{a}(p)\| > a_{\mathrm{QIC}}. \tag{4.58}$$

Assume that the optimal interference canceler satisfies constraint (4.57), i.e., that $\|\mathbf{a}_{\mathrm{opt}}(p)\| < a_{\mathrm{QIC}}$. Then, (4.57) also limits the mismatch, since $\|\mathbf{m}(p)\| = \|\mathbf{a}(p) - \mathbf{a}_{\mathrm{opt}}(p)\| < 2a_{\mathrm{QIC}}$.

The smaller a_{QIC} is, the lower the target and interference signal suppressions are. Appropriate values of a_{QIC} depend on the position of the sources and on the microphone arrangement. Thus, one must determine the smallest a_{QIC} that does not impair the interference signal suppression experimentally.

4.5 Experiments

In this section, we evaluate the performance in terms of SIR improvement of adaptive beamforming algorithms in the car environment. The following algorithms are compared:

- *NLMS.* The adaptation is performed according to (4.7) without double-talk detector.

---

[4] The argument p is omitted to simplify the notations.

- *ILMS.* The adaptation is performed according to (4.13) with step-size μ_0 and without double-talk detector. The stability threshold $\mu_{\max}$ is set to $\mu_{\max} = \frac{1}{2}$.
- *DTD-NLMS.* Adaptation of the interference canceler is performed according to (4.7) when the speech activity of the interferer is detected and the target is silent. The double-talk detection (DTD) is obtained by segmenting the recorded target source signal manually. In the case of the RGSC used with the four-element compact array mounted in the rear-view mirror, the blocking matrix is adapted when speech activity of the target is detected and the interferer is silent.

For these three algorithms, the adaptation is stopped if the short-term input signal level is below a fixed threshold. This threshold is determined by measuring the microphone signal energy during the first speech-free 200 ms. This rudimentary speech activity detection does not allow to distinguish between speech of the target (the driver) and speech of the interferer (the codriver). It is just useful to avoid divergence of the filter coefficients when the input signals consist of background noise only. The three algorithms are implemented with the quadratic inequality constraint in (4.58).

The performance measures

The performance measures have been defined in Sect. 2.4. The signal powers are estimated with their instantaneous estimates. In Figs. 4.2–4.5, the target signal level reduction and the interference signal level reduction are shown after averaging over a sliding window of length 30 ms.

Source signals

The online beamformer performance is obtained on real recordings performed with two male speakers. The driver utters a sequence of digits in German:

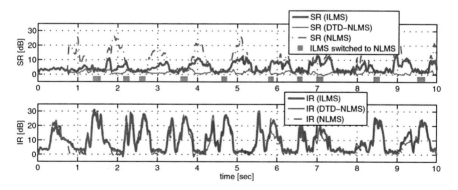

Fig. 4.2. Online performances with the microphone array mounted in the mirror (no background noise). The NLMS step-size is set to $\mu_{\mathrm{NLMS}} = 0.4$

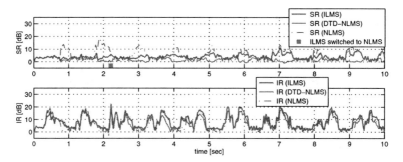

Fig. 4.3. Online performances with the four-element compact array mounted in the rear-view mirror, with road noise. The NLMS step-size is set to $\mu_{\text{NLMS}} = 0.3$

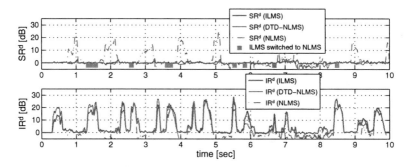

Fig. 4.4. Online performances with the two-element distributed array mounted on the car ceiling (no background noise)

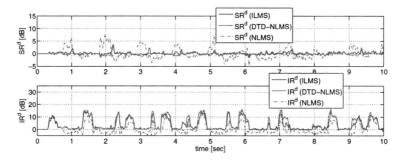

Fig. 4.5. Online performances with the two-element distributed array mounted on the car ceiling, with road noise. The step-size μ_{NLMS} is set to $\mu_{\text{NLMS}} = 0.05$

"*eins, zwei, ..., zehn*" ("one, two, ...,ten"), the codriver the digits from "*elf*" ("eleven") to "*neunzehn*" ("nineteen"). For the first half of the input signals, the recordings are interleaved so that the digit of one speaker falls mainly in the pause of the other one. By contrast, for the second part, the two speakers

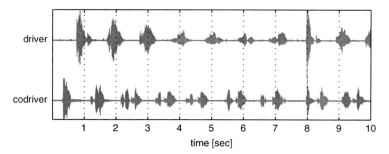

Fig. 4.6. Source signals

speak more or less simultaneously, as may be seen in Fig. 4.6. The signals are sampled at $f_s = 16\,\text{kHz}$. First, we consider the four-element compact array mounted in the rear-view mirror. Then, experiments are conducted with the two-element distributed array mounted on the car ceiling. (We refer to Appendix A for more details on the experimental setups.)

4.5.1 Experiments with the Four-Element Compact Array Mounted in the Rear-View Mirror

In our experiments, the four-element compact array mounted in the rear-view mirror is correctly steered to the target direction $\theta = 20°$. The uncontrolled NLMS and ILMS algorithms are used to adapt the interference canceler embedded in the GSC structure with the blocking matrix (3.29). The delay in the fixed beamformer path is $D = L/2 = 128$.

The controlled DTD-NLMS is used with the RGSC beamformer proposed by Hoshuyama et al. [53] (see also Sect. C.1 in Appendix C for more details). The RGSC has an adaptive blocking matrix (C.3) with M outputs and filters $\mathbf{b}_m$ of length L. The blocking matrix is adapted when the target is active and the interferer is silent with the NLMS algorithm (C.5) as described in Appendix C.

Determining the constants μ_{NLMS}, μ_0, and a_{QIC}

To determine the adaptation constants, a two-second recording carried out with an artificial head positioned on the codriver seat is used. The training signal is a female voice saying *"Ich bin Rudolf Ranick hier vom FTZ"* ("I am Rudolf Ranick here from FTZ"). The constants are tuned without double-talk, i.e., the artificial head of the codriver is the only active source and the microphone signals are those of the interferer, i.e., $\mathbf{x}(p) = \mathbf{n}(p)$. Optionally, road noise recorded at $100\,\text{km}\,\text{h}^{-1}$ is added to the microphone signals, which may be written as $\mathbf{x}(p) = \mathbf{n}(p) + \mathbf{n}^{(\text{road})}(p)$ in this case. This background noise exhibits a signal-to-noise ratio (SNR) of about $10\,\text{dB}$ with respect to codriver speech.

To determine the appropriate values for μ_{NLMS}, μ_0 and a_{QIC}, we define the quality measure Q as the interference signal level reduction estimated over the whole two-second recording, i.e.,

$$Q \triangleq \frac{\frac{1}{M} \sum_{m=1}^{M} \sum_{p=1}^{2f_s} n_m^2(p)}{\sum_{p=1}^{2f_s} y^2(p)}. \tag{4.59}$$

Thus, the initial convergence phase of $\mathbf{a}(p)$ is included in Q. Note that we are interested in the reduction of the passenger speech level, which is measured by Q (as opposed to the reduction of the road noise level). The target signal cancelation problem is not considered for now. As we will see, Q decreases in the presence of background noise since some degrees of freedom of the interference canceler are allocated to the attenuation of the background noise signal rather than to the suppression of the interferer signal.

First, the step-size is examined. The quadratic constraint (4.57) is left out, i.e., $a_{\text{QIC}} = +\infty$ in (4.58). Q is depicted in Fig. 4.7a as a function of μ_0 and μ_{NLMS}. Note that $Q(\mu_0)$ exhibits two local maxima. This may be explained as follows: For a large step-size μ_0, the stability condition (4.11) is not fulfilled and the algorithm switches to the NLMS adaptation. Thus, $Q(\mu_0) \approx Q(\mu_{\text{NLMS}})$ for $\mu_0, \mu_{\text{NLMS}} > 0.1$. Since most of the adaptation should occur with the pseudooptimal step-size, we retain the first maximum of $Q(\mu_0)$ at $\mu_0 = 0.008$. Let us also give an interpretation of the curve $Q(\mu_0)$ in terms of the contraction factor $\alpha(p)$ introduced in (4.39) in Sect. 4.3.2. $Q(\mu_0)$ reaches its first maximum when the contraction factor $\alpha(p)$ is minimum. For $10^{-2} < \mu_0 < 10^{-1}$ corresponds to an overshoot phase: the contraction factor becomes negative and larger in magnitude, which decreases the speed of convergence.

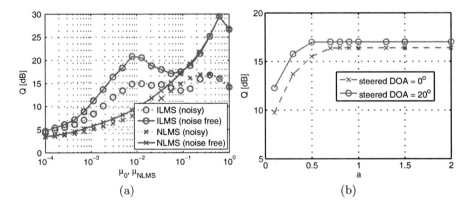

(a)   (b)

Fig. 4.7. (a) Q as a function of μ_0 and μ_{NLMS}, for $a_{\text{QIC}} = +\infty$. (b) Q as a function of a_{QIC}, obtained with NLMS and the step-size $\mu_{\text{NLMS}} = 0.2$ and noisy input signals. These results are obtained with the four-element compact array mounted in the rear-view mirror

The algorithm switches to the NLMS adaptation when the contraction factor becomes too large (which would cause instability).

The noisy case deserves an additional observation. The maximum value of $Q(\mu_{\text{NLMS}})$ indicates that the NLMS algorithm is able to attain a higher interference signal suppression than ILMS on noise-free input signals. However, both algorithms exhibit similar performance with background noise. Interestingly, the background noise does not seem to influence the optimum μ_0, which indicates the robustness of the ILMS algorithm against varying noise conditions.

Second, we use the noisy input signals to determine the smallest a_{QIC} that does not impair the interference signal level reduction. The value of Q as a function of a_{QIC} is depicted in Fig. 4.7b. Note that the norm of the optimal interference canceler coefficients $\|\mathbf{a}_{\text{opt}}\|$ depends on the proper array steering: Steering errors may cause (1) larger interference signal levels in the target reference $x_0(p)$ and (2) weaker interference signal levels in the interference references $x_{B,m}(p)$. Then the optimal a_{QIC} may depend on steering errors. That is why the results in Fig. 4.7b are shown for the correct steering direction ($20°$) and for an erroneous steering direction ($0°$). It shows that for $a_{\text{QIC}} > 0.8$, the interference signal level reduction is barely impaired for both steering directions. The parameter values $\mu_0 = 0.008$ and $a_{\text{QIC}} = 0.8$ are used in the following. The algorithm equation and parameters for the four-element compact array mounted in the rear-view mirror are summarized in Table 4.1.

Online performance

Figure 4.2 presents the online performance with noise-free signals and Fig. 4.3 those with road background noise.

Except for the fact that the presence of background noise decreases the target signal level reduction $SR(p)$ and the interference signal level reduction[5] $IR(p)$, both figures have similar characteristics:

Table 4.1. DTD-NLMS and ILMS algorithm short reference table with the parameter settings for the compact four-element compact array mounted in the rear-view mirror shown in Fig. A.2. The filter length is set to $L = 256$ at $f_{\text{s}} = 16\,\text{kHz}$

	DTD-NLMS	ILMS
structure	RGSC (Fig. C.1)	GSC (Fig. 3.1)
section	4.1	4.2
equation	(4.7)	(4.13)
step-size	$\mu_{\text{NLMS}} = 0.4$ (no noise)	$\mu_0 = 0.008$
	$\mu_{\text{NLMS}} = 0.3$ (noisy)	$\mu_{\text{max}} = \frac{1}{2}$
QIC	$a_{\text{QIC}} = 0.8$	

---

[5] Note that the interference signal level reduction does not measure the attenuation of the background noise signal but only that of the codriver speech.

- All three algorithms provide similar reductions of the interference signal level IR(p).
- The uncontrolled NLMS reduces the target and interferer signal levels similarly and therefore does not provide any SIR improvement. This confirms the result of Sect. 4.3.2, namely that NLMS converges and diverges with the same speed.
- The target signal level reduction SR(p) produced by ILMS increases slowly. SR(p) remains smaller than for the uncontrolled NLMS. This reflects the result of Sect. 4.3.2, namely that ILMS converges faster than it diverges. Nevertheless, SR(p) may attain 10 dB when the driver speaks alone and is similar to that of NLMS at the end of the recording. The periods of time where ILMS is switched to NLMS, i.e., when the stability condition (4.11) is not fulfilled, are indicated below the SR(p) curves. As shown in Figs. 4.2 and 4.6, instability is detected only when the target is weak relative to the interferer, as predicted in Sect. 4.3.2. This occurs less frequently with background noise: The diffuse road noise cannot be completely canceled and $\|\mathbf{y}(p)\|^2$ does not become very small, preventing large values of $1/\|\mathbf{y}(p)\|^2$ in (4.11).
- The controlled DTD-NLMS method is the only one that guarantees a low target signal cancelation during the whole recording and offers the best SIR improvement. This shows that it is necessary to interrupt the adaptation during target activity to prevent target signal cancelation with the four-element compact array mounted in the rear-view mirror.

To examine the distortion of the desired signal for each algorithm, we consider the signal level reduction as a function of the frequency.[6] The distortion curves are plotted in Fig. 4.8. The figure is in accordance with the observation above: the ILMS algorithm causes a significantly higher distortion than the DTD-NLMS.

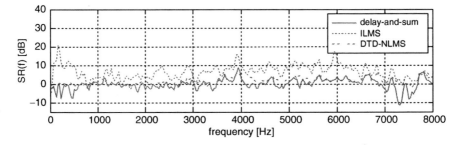

Fig. 4.8. Signal level reduction as a function of the frequency for the four-element compact array mounted in the rear-view mirror (no background noise)

---

[6] The power spectral density (PSD) was estimated using Welch's periodogram method with Hanning-weighted time frames of length 512. The PSD was then averaged over the whole signal length.

4.5.2 Experiment with the Two-Element Distributed Array Mounted on the Car Ceiling

The two-element distributed array mounted on the car ceiling is associated with the causal AIC structure described in Sect. 3.3.2. The experimental setup is depicted in Fig. A.3.

Determining the constants μ_{NLMS}, μ_0, and a_{QIC}

The same procedure as in Sect. 4.5.1 is applied. The quality measure Q is the reduction of the interference signal energy evaluated with respect to the first microphone $x_1(p)$:

$$Q \triangleq \frac{\sum_{p=1}^{2f_s} n_1^2(p)}{\sum_{p=1}^{2f_s} y^2(p)} \qquad (4.60)$$

Q is shown as a function of μ_0 and μ_{NLMS} in Fig. 4.9a. $Q(\mu_0)$ exhibit two local maxima for the same reasons as given in Sect. 4.5.1. The relevant one is the first, at $\mu_0 = 0.005$. The second maximum of $Q(\mu_0)$ for $\mu_0 > 0.1$ appears because of switching to NLMS. Note that the step-size μ_0 which maximizes Q is almost independent of the background noise. In this respect, the robustness of the ILMS adaptation is remarkable. By contrast, the NLMS step-size μ_{NLMS} needs to be adjusted to the noise level: the step-size is set to $\mu_{NLMS} = 0.2$ in noise-free conditions, and to $\mu_{NLMS} = 0.05$ in noisy conditions. Regarding the upper bound a_{QIC} for the quadratic inequality constraint $\|\mathbf{a}\| < a_{QIC}$, it appears in Fig. 4.9b that the conservative $a_{QIC} = 0.5$ does not impair the reduction of the interference signal level. We retain $\mu_0 = 0.005$ and $a_{QIC} = 0.5$

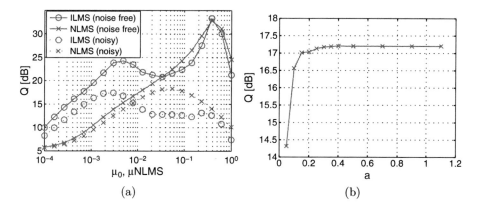

(a)   (b)

Fig. 4.9. (a) Q as a function of μ_0 and μ_{NLMS}, for $a_{QIC} = +\infty$. (b) Q as a function of a_{QIC}, obtained on noisy input signals with NLMS and the step-size $\mu_{NLMS} = 0.05$. Results obtained with the two-element distributed array mounted on the car ceiling

Table 4.2. DTD-NLMS and ILMS algorithm short reference table with the parameter settings for the two-element distributed array mounted on the car ceiling shown in Fig. A.3. The filter length is set to $L = 256$ at $f_s = 16\,\text{kHz}$

	DTD-NLMS	ILMS
structure	interference canceler (Fig. 3.2)	
section	4.1	4.2
equation	(4.7)	(4.13)
step-size	$\mu_{\text{NLMS}} = 0.1$ (no noise)	$\mu_0 = 0.005$
	$\mu_{\text{NLMS}} = 0.05$ (noisy)	$\mu_{\text{max}} = \frac{1}{2}$
QIC	$a_{\text{QIC}} = 0.5$	

for further experiments. The algorithm equation and parameter settings are summarized in Table 4.2.

Online performance

The online performance with noise-free signals is shown in Fig. 4.4, and with road background noise in Fig. 4.5.

It can be observed that the uncontrolled NLMS leads to a significant target signal level reduction, even though this reduction is not as large as with the four-element compact array mounted in the rear-view mirror. Although the prior information about the source position is used at the physical level with directive microphones and with a conservative constraint $\|\mathbf{a}(p)\| < 0.5$, target signal cancelation nevertheless occurs. This influences the interference signal suppression negatively, since the degrees of freedom that are allocated to the target signal cancelation are not available for the suppression of the interference signal.

By contrast, the ILMS method does not lead to a noticeable target signal level reduction. This result differs from the one obtained with the four-element compact array mounted in the rear-view mirror and may be explained as follows:

- The long-term average interferer-to-signal ratio at the microphone that is oriented to the codriver is about 6 dB. This helps to limit the energy inversion effect described in Sect. 3.4.1.
- The number of spatial degrees of freedom allows only a single spatial zero to be adaptively placed. This prevents the driver and codriver speech to be simultaneously canceled.
- A causality constraint is set on the interference canceler, which also helps to decrease target signal level reduction.

The reduction of the desired signal level as a function of the frequency is shown in Fig. 4.10 to illustrate the distortion caused by the ILMS and DTD-NLMS algorithms.

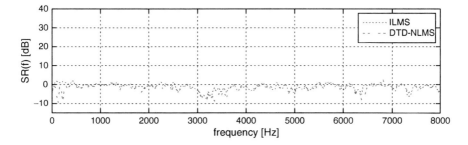

Fig. 4.10. Reduction of the desired signal level as a function of the frequency for the ILMS and DTD-NLMS algorithms with the two-element distributed array mounted on the car ceiling (no background noise). The PSD is averaged over the whole signal length

As predicted in Sect. 4.3.2, ILMS switches in noise-free conditions to NLMS only if the input SIR is low. With background noise, the output signal power is such that instability is never detected.

The fastest convergence is attained with the controlled DTD-NLMS algorithm. The interference signal level reduction is somewhat limited during double-talk since no adaptation occurs when the target is active, but it is still comparable to that of the ILMS.

4.6 Summary and Conclusion

In this chapter, we have introduced adaptive LCMV beamforming algorithms and built upon the widely used NLMS algorithm. An important parameter of the NLMS algorithm is the step-size. A large step-size is desirable since it allows a rapid tracking of the spectral changes. However, a large step-size also leads to a significant target signal suppression if the input SIR is large or during double-talk. We addressed this problem by using a pseudooptimal step-size. This leads to the ILMS algorithm with implicit adaptation control. On the theoretical side, ILMS has been shown to converge faster and diverge slower than the NLMS algorithm without adaptation control. It was also shown how the associated stability condition acts as a target silence detector.

To further increase the robustness against target signal cancelation, we integrated a quadratic inequality constraint. The performance in terms of SIR improvement of the developed ILMS algorithm was then experimentally studied and compared to that of the traditional NLMS algorithm. The theoretical results on the transient behavior and the stability could be confirmed. Moreover, it appears that the ILMS parameter μ_0 does not require to be adjusted to the background noise level. This feature makes the ILMS very attractive for automotive applications. The SIR improvement provided by

ILMS outperforms that of the uncontrolled NLMS algorithm. With the two-element distributed array mounted on the car ceiling, the performance of the ILMS algorithm is comparable to that of the DTD-controlled NLMS, while not requiring any external adaptation control. With this setup, the ILMS algorithm is very robust against target leakage and has a high practical relevance. On the other hand, for the four-element compact array mounted in the rear-view mirror, the adaptation control provided by the ILMS algorithm is not sufficient to prevent the target signal.

To summarize, the most important results of this chapter are:

- In contrast to the NLMS algorithm, the step-size in the ILMS algorithm is automatically adjusted to the input SIR. Moreover, it also adjusts to the background noise level, an important feature in automotive applications.
- Used in conjunction with the distributed microphone arrangement and an AIC, this implicit adaptation control seems to be sufficient. However, in the general case of GSC beamformers, the ILMS algorithm alone does not prevent target leakage and an external adaptation control is still necessary.

5

Second-Order Statistics Blind Source Separation

The adaptive LCMV beamforming methods discussed in Chap. 4 are based on the constrained minimization of the output power as an optimization criterion. In practice, their performance is subjected to two contradictory constraints. On the one hand, the microphones should be placed as close as possible to the desired source for a good acoustic capture. On the other hand, any leakage of the desired signal in the interferer reference results in a cancelation of the desired signal at the beamformer output and in a poor SIR improvement.

This fundamental limitation may be overcome with source *separation* methods. In contrast to power-based adaptive beamforming where only the target signal estimate

$$y(p) = \sum_{m=1}^{M} \mathbf{w}_m^{\mathrm{T}}(p)\mathbf{x}_m(p) \tag{5.1}$$

is considered as system output, source separation methods include N source signal estimates $y_n(p)$ given by (2.23):

$$y_n(p) = \sum_{m=1}^{M} \mathbf{w}_{nm}^{\mathrm{T}}(p)\mathbf{x}_m(p) \tag{2.23}$$

for $n = 1, \ldots, N$. No formal distinction is made between the desired signal and the interference signals since both are recovered as algorithm output. These separation methods do not require any adaptation control, they are thus referred to as unsupervised, or *blind*[1] source separation methods (BSS).

A simple but illustrative derivation of these BSS principles was proposed by Van Gerven et al. with the SAD algorithm [38]. Their starting point is

---

[1] The term "blind" indicates that no a priori known training sequence is necessary to adapt the separation filters [46]. By contrast, the power-based LMS in Chap. 4 is supervised (nonblind) and necessitates an a priori known training sequence. This sequence consists of the desired signal during its silences and is zero.

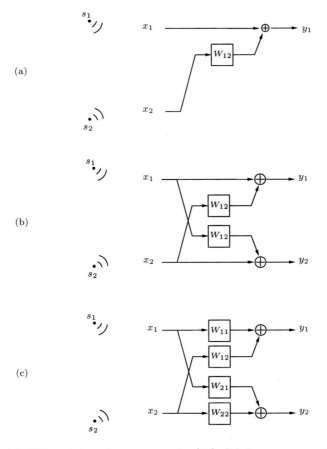

Fig. 5.1. (a) Widrow's interference canceler [92]. (b) Source separation structure
proposed by Van Gerven [38]. (c) Complete separation structure

the LMS algorithm (4.3) with the two-source two-sensor scenario shown in
Fig. 5.1a. The adaptive interference canceler is denoted by $\mathbf{w}_{12}(p)$ and the
output signal is given by $y_1(p) = x_1(p) + \mathbf{w}_{12}^{\mathrm{T}}(p)\mathbf{x}_2(p)$. The LMS algorithm
for $\mathbf{w}_{12}(p)$ can be written using (4.3) as

$$\mathbf{w}_{12}(p+1) = \mathbf{w}_{12}(p) - \mu y_1(p)\mathbf{x}_2(p). \tag{5.2}$$

Van Gerven takes into account the fact that the interferer reference signal
$\mathbf{x}_2(p)$ is contaminated by target signal components (the target leakage). He
proposes replacing $\mathbf{x}_2(p)$ in (5.2) by a target-free signal given by $y_2(p) =
x_2(p) + \mathbf{w}_{21}^{\mathrm{T}}(p)\mathbf{x}_1(p)$ and obtains two update rules for $\mathbf{w}_{12}(p)$ and $\mathbf{w}_{21}(p)$:

$$\begin{cases} \mathbf{w}_{12}(p+1) = \mathbf{w}_{12}(p) - \mu y_1(p)\mathbf{y}_2(p), \\ \mathbf{w}_{21}(p+1) = \mathbf{w}_{21}(p) - \mu y_2(p)\mathbf{y}_1(p). \end{cases} \tag{5.3}$$

The corresponding feedforward filter structure proposed by Van Gerven is shown in Fig. 5.1b. This adaptive MIMO provides an estimate for the target signal $\mathbf{y}_1(p)$ but also an estimate $\mathbf{y}_2(p)$ for the interferer signal as a by-product. When (5.3) has converged in the mean, $\mathbf{y}_1(p)$ and $\mathbf{y}_2(p)$ should be uncorrelated. Hence, Van Gerven termed the algorithm in (5.3) the symmetric adaptive decorrelation (SAD) algorithm [38].

Unfortunately, (5.3) yields severe limitations. Like the LMS algorithm, (5.3) is not normalized and is predisposed to instability if the input signals have fluctuating powers, as it is the case for speech signals. Moreover, with (5.3), the zero-tap filter coefficients $w_{12,0}(p)$ and $w_{21,0}(p)$ are such that $w_{12,0}(p) = w_{21,0}(p)$ for all p. Therefore, the practical use of the SAD algorithm (5.3) is limited.

BSS algorithms are based on the assumption that the source signals are mutually independent and adjust the filters $\mathbf{w}_{nm}$ to minimize a certain dependence measure. In this book, the emphasis is placed on second-order statistics BSS (SOS-BSS). That is, the dependence measure is based on the second-order statistics of the source signals. Other dependence measures based on higher-order statistics exist and yield different BSS algorithms [56].

This chapter focuses on *offline* separation methods in the time domain, for which the entire observed signal $\mathbf{x}(p), p = 1, \ldots, T$ is available at the time of processing. The presentation makes use of Sylvester matrices and is derived from Buchner et al. [18].

This chapter is organized as follows: Section 5.1 states the problem of blind source separation and defines the notations used. In Sect. 5.2, the role of the nonstationarity is explained and a separation cost function is derived using the mutual information of nonstationary Gaussian signals. In Sects. 5.3 and 5.4, the gradient of this cost function is given and we introduce the natural gradient that is more efficient. Unfortunately, the natural gradient applies only to "square" systems that have as many sources as microphones. In Sect. 5.4, we propose a general approach to extend the natural gradient to nonsquare systems.

5.1 Problem and Notations

5.1.1 From a Scalar to a Convolutive Mixture Model

The scalar model

Originally the problem of blind source separation was placed in the framework of instantaneous (or scalar) linear mixtures [59]. In this framework, it is often termed independent component analysis (ICA). In the instantaneous model, N source signals $s_1(p), \ldots, s_N(p)$ propagate instantaneously to M sensors $x_1(p), \ldots, x_M(p)$. That is, the source–sensor relationship can be described using an $M \times N$ matrix $\mathbf{H}$ as follows:

$$\mathbf{x}(p) \triangleq \mathbf{H}\mathbf{s}(p), \tag{5.4}$$

$$\text{with } \mathbf{s}(p) \triangleq (s_1(p), \dots, s_N(p))^{\mathrm{T}} \tag{5.5}$$

$$\text{and } \mathbf{x}(p) \triangleq (x_1(p), \dots, x_M(p))^{\mathrm{T}}. \tag{5.6}$$

The matrix $\mathbf{H}$ is called the *mixing* matrix. Roughly speaking, the objective of BSS is to isolate the image of one source in each component of the output vector $\mathbf{y}(p) = (y_1(p), \dots, y_N(p))^{\mathrm{T}}$ defined as

$$\mathbf{y}(p) \triangleq \mathbf{W}\mathbf{x}(p). \tag{5.7}$$

The $N \times M$ matrix $\mathbf{W}$ is called a *separation* matrix. Note that it is not necessary to identify $\mathbf{H}^{-1}$ (if it is a square matrix) since recovering the sources up to permutations and scaling is sufficient. In BSS, the goal is to blindly identify a separation matrix $\mathbf{W}$ such that the matrix $\mathbf{W}\mathbf{H}$ has only one non-zero entry in each column and each row. Such a separation matrix is called a *separating* matrix.[2] Although the instantaneous model is of limited practical relevance for acoustic applications, its role is fundamental in the development and understanding of BSS algorithms. In the following, this model is extended to the convolutive case by introducing specific notations [20].

The convolutive model

First, we generalize the scalar output components $y_n(p)$ in (5.7) to vectors $\mathbf{y}_n(p)$ of length L, as follows:

$$\mathbf{y}_n(p) \triangleq (y_n(p), y_n(p-1), \dots, y_n(p-L+1))^{\mathrm{T}}. \tag{5.8}$$

The (n, m)-element of the separation matrix, w_{nm}, becomes a Sylvester matrix $\mathbf{W}_{nm}$ of size $L \times 2L - 1$ which is defined as

$$\mathbf{W}_{nm} \triangleq \begin{pmatrix} \mathbf{w}_{nm}^{\mathrm{T}} & 0 & \cdots & 0 \\ 0 & \mathbf{w}_{nm}^{\mathrm{T}} & \ddots & \vdots \\ \vdots & \ddots & \ddots & 0 \\ 0 & \cdots & 0 & \mathbf{w}_{nm}^{\mathrm{T}} \end{pmatrix}, \tag{5.9}$$

$$\mathbf{w}_{nm}^{\mathrm{T}} \triangleq (w_{nm,0}, \dots, w_{nm,L-1}). \tag{5.10}$$

The entire MIMO separation system is given by the block Sylvester matrix $\mathbf{W}$:

$$\mathbf{W} \triangleq \begin{bmatrix} \mathbf{W}_{11} & \cdots & \mathbf{W}_{1M} \\ \vdots & \ddots & \vdots \\ \mathbf{W}_{N1} & \cdots & \mathbf{W}_{NM} \end{bmatrix}. \tag{5.11}$$

---

[2] Note that it is a priori assumed that there exists a separating matrix that does not depend on a particular source realization [23].

The input data $x_m(p)$ are stacked in a vector $\mathbf{x}(p)$ that is now redefined[3] as

$$\mathbf{x}(p) \triangleq \left(\mathbf{x}_1^{\mathrm{T}}(p), \ldots, \mathbf{x}_M^{\mathrm{T}}(p)\right)^{\mathrm{T}}, \tag{5.12}$$

$$\text{with } \mathbf{x}_m(p) \triangleq (x_m(p), \ldots, x_m(p - 2L + 2))^{\mathrm{T}} \quad \text{for } m = 1, \ldots, M. \tag{5.13}$$

Then the MIMO input–output relationship in (2.22) can be compactly written as

$$\mathbf{y}(p) = \mathbf{W}\mathbf{x}(p). \tag{5.14}$$

Up to the fact that the input vector $\mathbf{x}(p)$ now has size $M(2L-1) \times 1$ and that the output signal $y(p)$ becomes a vector $\mathbf{y}(p)$, (5.14) is simply a compact version of (2.23) and thus the Sylvester notation is consistent with the notations of the previous chapters. In the case $M = N = 2$, the input–output structure is shown in Fig. 5.1c.

Similarly, the mixing equation can be written as

$$\mathbf{x}(p) \triangleq \mathbf{H}\mathbf{s}(p). \tag{5.15}$$

Each scalar mixing coefficient h_{ij} of the matrix $\mathbf{H}$ in (5.4) is generalized to a mixing filter, which is represented by the $(2L - 1) \times (L_{\mathrm{m}} + 2L - 2)$ Sylvester matrix $\mathbf{H}_{ij}$ as follows:

$$\mathbf{H}_{ij} \triangleq \begin{pmatrix} \mathbf{h}_{ij}^{\mathrm{T}} & 0 & \cdots & 0 \\ 0 & \mathbf{h}_{ij}^{\mathrm{T}} & \ddots & \vdots \\ \vdots & \ddots & \ddots & 0 \\ 0 & \cdots & 0 & \mathbf{h}_{ij}^{\mathrm{T}} \end{pmatrix}, \tag{5.16}$$

$$\mathbf{h}_{ij}^{\mathrm{T}} \triangleq (h_{ij,0}, \ldots, h_{ij,L_{\mathrm{m}}-1}). \tag{5.17}$$

L_{m} denotes the length of the mixing channels. The whole MIMO mixing system is given by the block Sylvester matrix $\mathbf{H}$ of size $M(2L - 1) \times N(L_{\mathrm{m}} + 2L - 2)$:

$$\mathbf{H} \triangleq \begin{bmatrix} \mathbf{H}_{11} & \cdots & \mathbf{H}_{1N} \\ \vdots & \ddots & \vdots \\ \mathbf{H}_{M1} & \cdots & \mathbf{H}_{MN} \end{bmatrix}. \tag{5.18}$$

The source vector $\mathbf{s}(p)$ in (5.15) is defined as

$$\mathbf{s}(p) \triangleq \left(\mathbf{s}_1^{\mathrm{T}}(p), \ldots, \mathbf{s}_N^{\mathrm{T}}(p)\right)^{\mathrm{T}}, \tag{5.19}$$

$$\text{with } \mathbf{s}_n(p) \triangleq (s_n(p), \ldots, s_n(p - 2L - L_{\mathrm{m}} + 3))^{\mathrm{T}} \text{ for } n = 1, \ldots, N. \tag{5.20}$$

---

[3] In Chaps. 2–4, the vector $\mathbf{x}(p)$ was defined with length ML. In source separation, this vector has length $M(2L - 1)$.

This notation with Sylvester matrices is appealing because convolutive mixtures appear formally as instantaneous mixtures, which can be seen by comparing (5.4) with (5.15) or (5.7) with (5.14). Using this type of notation, instantaneous separation methods have been proposed for separating convolutive mixtures [70]. However, for acoustic signal processing, several tens or hundreds of filter taps are necessary. This makes the dimension of the mixing prohibitively high. Nevertheless, such a notation allows to apply some results which have been derived for instantaneous BSS.

Finally, let us define the global convolutive system $\mathbf{C} \triangleq \mathbf{WH}$. The $NL \times N(L_m + 2L - 2)$ matrix $\mathbf{C}$ contains N^2 submatrices $\mathbf{C}_{nn'}, n, n' = 1, \ldots, N$ of size $L \times L_m + 2L - 2$:

$$\mathbf{C} = \begin{bmatrix} \mathbf{C}_{11} & \cdots & \mathbf{C}_{1N} \\ \vdots & \ddots & \vdots \\ \mathbf{C}_{N1} & \cdots & \mathbf{C}_{NN} \end{bmatrix}. \tag{5.21}$$

Each submatrix $\mathbf{C}_{nn'}$ represents the channel that relates the source $s_{n'}(p)$ to the output $y_n(p)$.

5.1.2 Separation Ambiguities

Generally, blind separation criteria are based on the mutual independence of the sources. Minimizing an independence criterion does not allow the separating matrix $\mathbf{W}$ to be uniquely determined. In the following, we describe the permutation and scaling ambiguities in the convolutive case. Let us assume that the output signals are mutually independent. Since permutations of the outputs leave their mutual independence unchanged, $\mathbf{y}_m(p)$ is not necessarily an estimate of the source signal $\mathbf{s}_m(p)$. In fact, the ordering of the source signals is solely a matter of notation and has no physical relevance. Therefore, the output vector $\mathbf{y}(p)$ should provide an estimate of the source vector $\mathbf{s}(p)$ up to arbitrary permutations. In terms of the global system $\mathbf{C}$, this means that we do not want to find $\mathbf{W}$ such that $\mathbf{C}$ is block-diagonal. Rather we want to find $\mathbf{W}$ such that each row and each column of $\mathbf{C}$ has only one nonzero submatrix.

Likewise, individual filtering of each source signal leaves the mutual independence of the sources unchanged. Therefore, the nonzero submatrices in $\mathbf{C}$ may contain arbitrary filter coefficients. This filtering indeterminacy may be more or less severe depending on the filter length L: If L is greater than the lower bound in (2.55), i.e., if $L > \left\lceil \frac{(L_m - 1)(N-1) + 1}{M - N + 1} \right\rceil$, the unconstrained degrees of freedom result in a true arbitrary filtering of the source signals. If L equals the lower bound, this arbitrary filtering reduces to an arbitrary scaling [19]. To neutralize this ambiguity, various normalizations may be adopted.

For example, it may be useful to constrain the diagonal filters $\mathbf{w}_{nn}$ to unit responses. However, depending on the length L_m of the mixing channels, such a constraint generally reduces the achievable separation performance, as we have shown in Sect. 2.3.2.

5.2 Nonstationarity and Source Separation

5.2.1 The Insufficiency of Decorrelation

In most instantaneous BSS approaches, the source signals $s_n(p), n = 1, \ldots, N$ are modeled as realizations of stationary stochastic processes [23]. With this model, the blind separation of Gaussian sources is not possible: As the mutual independence and the decorrelation of Gaussian sources are equivalent, the independence constraints become the following decorrelation constraints:

$$\mathbf{R}_{\mathbf{y}_n \mathbf{y}_{n'}} \triangleq \mathbf{E}\left\{\mathbf{y}_n(p)\mathbf{y}_{n'}^{\mathrm{T}}(p)\right\} = \mathbf{0} \quad \forall n \neq n'. \tag{5.22}$$

Suppose $\mathbf{w}_{nn} = \boldsymbol{\delta}_0$ for $n = 1, \ldots, N$, which neutralizes the filtering ambiguity. On the one hand, the independence criterion (5.22) provides $(2L - 1)(N^2 - N)/2$ constraints. On the other hand, we have $(N^2 - N)L$ unknown filter coefficients $w_{nm,k}$. Therefore, a continuum of solutions with dimension $(N^2 - N)/2$ arises, which is "larger" than the discrete set of permutation ambiguities. This shows that Gaussian sources cannot be separated if we use the stationary signal model. This also reveals that the decorrelation constraints (5.22) are not sufficient to separate stationary source signals, whether Gaussian or not. This is the motivation for an alternative signal model. Instead of stationary source realizations, it is proposed to examine nonstationary source signals. Although the nonstationarity is often an obstacle in adaptive signal processing, it may be advantageously exploited in BSS. If we assume nonstationary source signals, we may produce $K(2L - 1)(N^2 - N)/2$ equations from (5.22) by setting $p = t_1, \ldots, t_K$, that is,

$$\mathbf{E}\left\{\mathbf{y}_n(t_k)\mathbf{y}_{n'}^{\mathrm{T}}(t_k)\right\} = \mathbf{0}, \quad k = 1, \ldots, K, \quad \forall n \neq n'. \tag{5.23}$$

For $K \geq 2$, the collection of constraints is sufficient to identify appropriate separating filters. These constraints correspond to the joint block diagonalization of the output correlation matrices $\mathbf{R}_{\mathbf{yy}}(t_k), k = 1, \ldots, K$, where $\mathbf{R}_{\mathbf{yy}}(p)$ is defined as

$$\mathbf{R}_{\mathbf{yy}}(p) \triangleq \mathbf{E}\left\{\mathbf{y}(p)\mathbf{y}^{\mathrm{T}}(p)\right\}. \tag{5.24}$$

Let us introduce the matrix operator $\mathrm{boff}(\mathbf{R})$ that sets the diagonal $L \times L$ submatrices of $\mathbf{R}$ to zero. For a $K_1 L \times K_2 L$ matrix $\mathbf{R}$ consisting of submatrices $\mathbf{R}_{ij}, i = 1, \ldots, K_1; j = 1, \ldots, K_2$ of size $L \times L$, $\mathrm{boff}(\mathbf{R})$ is defined as

$$\text{boff}\begin{bmatrix} \mathbf{R}_{11} & \cdots & \mathbf{R}_{1K_2} \\ \vdots & \ddots & \vdots \\ \mathbf{R}_{K_11} & \cdots & \mathbf{R}_{K_1K_2} \end{bmatrix} \triangleq \begin{bmatrix} \mathbf{0} & \mathbf{R}_{12} & \cdots & & \mathbf{R}_{1K_2} \\ \mathbf{R}_{21} & \ddots & & \ddots & \vdots \\ \vdots & \ddots & & & \mathbf{R}_{(K_1-1)K_2} \\ \mathbf{R}_{K_11} & \cdots & \mathbf{R}_{K_1(K_2-1)} & & \mathbf{0} \end{bmatrix}. \quad (5.25)$$

Reformulating (5.23), we may resolve the separation problem by finding $\mathbf{W}$ such that

$$\text{boff}(\mathbf{R}_{\mathbf{yy}}(t_k)) = \mathbf{0} \quad k = 1,\ldots,K, \quad (5.26)$$

since $\text{boff}(\mathbf{R}_{\mathbf{yy}}(t_k))$ contains all cross-correlation terms.

5.2.2 Nonstationarity-Based Cost Function

Gaussian mutual information

Various cost functions based on the joint block diagonalization of the correlation matrices $\mathbf{R}_{\mathbf{yy}}(t_k)$ have been introduced heuristically (see e.g.,[75]).

By contrast, the mutual information,[4] a natural measure of independence in information theory, gives us a rigorous foundation for deriving a separation

---

[4] Some definitions and properties from the information theory are essential [29]:

- The KULLBACK-LEIBLER divergence between two distributions

$$D(p_X, p_Y) \triangleq \int_u p_X(u) \log \frac{p_X(u)}{p_Y(u)} \mathrm{d}u$$

is not really a distance in topological terms. For example, it is not symmetrical, i.e., $D(p,q) \neq D(q,p)$ in general. However, it may be seen as a distance between two random variables. It is clear that $D(p,p) = 0$ for all probability density function (PDF) p. Moreover, $D(p,q)$ is always positive.
- $D(p_{(X,Y)}, p_X \cdot p_Y)$ is called the mutual information of X and Y, and is denoted by

$$I(X,Y) \triangleq D(p_{(X,Y)}, p_X \cdot p_Y) \quad (5.27)$$

$$= \int_{x,y} p_{(X,Y)}(x,y) \log \frac{p_{(X,Y)}(x,y)}{p_X(x) \cdot p_Y(y)} \mathrm{d}x\mathrm{d}y. \quad (5.28)$$

The mutual information is nonnegative, and a very natural measure of independence. The equivalence

$$X \text{ and } Y \text{ are independent} \Leftrightarrow I(X,Y) = 0$$

is fundamental. The entropy of X, defined as $H(X) \triangleq I(X,X)$, is related to the mutual information by

$$I(X,Y) = H(X) + H(Y) - H(X,Y). \quad (5.29)$$

cost function. To obtain a cost function from the mutual information, we must assume that if $\mathbf{W}$ is a separating matrix then the output signals have a particular PDF. Note that even if the output does not match the assumed p.d.f. when $\mathbf{W}$ is a separating matrix, the sources may be separated by minimizing the mutual information. In fact, the assumed p.d.f. should be simple enough to produce simple algorithms, while capturing the nonstationarity of the sources. Such a model may be obtained by considering that if $\mathbf{W}$ is a separating matrix then the outputs are realizations of mutually independent nonstationary Gaussian processes. In addition, it is assumed that the output signals have no time structure over blocks of length L, i.e., $\mathbf{y}_n(t_1)$ is assumed to be independent of $\mathbf{y}_n(t_2)$ for all n and all $t_1 \neq t_2$ such that $|t_1 - t_2| > L$. Again, we emphasize that this is a working assumption. The derived algorithms may be able to separate a larger class of source signals. Then, the mutual information of the entire output sequence[5] $\mathbf{y}(p), p = 1, \ldots, T$ is given by the sum of the mutual information in each block, that is,

$$I(\mathbf{y}) = \sum_{k=1}^{K} I\left(\mathbf{y}(kL)\right), \tag{5.30}$$

where $K = \lfloor T/L \rfloor$. To compute the output mutual information $I(\mathbf{y}(p))$ at time p, we use the Shannon entropy H [85] and the relation

$$I(\mathbf{y}(p)) = \sum_{n=1}^{N} H(\mathbf{y}_n(p)) - H(\mathbf{y}(p)). \tag{5.31}$$

Assuming that the output samples may be described as the realization of a stochastic process with a NL-variate normal distribution with correlation matrix $\mathbf{R_{yy}}(p)$, their entropy is given by [85]

$$H(\mathbf{y}(p)) = \frac{1}{2} \log\left((2\pi e)^{NL} \det \mathbf{R_{yy}}(p)\right). \tag{5.32}$$

Similarly, the entropy of $\mathbf{y}_n$ is given by $H(\mathbf{y}_n(p)) = \frac{1}{2} \log\left((2\pi e)^{L} \det \mathbf{R_{y_n y_n}}(p)\right)$, where $\mathbf{R_{y_n y_n}}(p) = \mathbf{E}\left\{\mathbf{y}_n(p)\mathbf{y}_n^{\mathrm{T}}(p)\right\}$. Then the mutual information $I(\mathbf{y}(p))$ in (5.31) is given by

$$I(\mathbf{y}(p)) = \frac{1}{2}\left(\sum_n \log \det \mathbf{R_{y_n y_n}}(p) - \log \det \mathbf{R_{yy}}(p)\right). \tag{5.33}$$

As may be seen from (5.33), the mutual information $I(\mathbf{y}(p))$ vanishes if $\mathbf{R_{yy}}(p)$ is block-diagonal, that is, if the output signals are spatially uncorrelated at

---

[5] For simplicity, the mutual information and the entropy are respectively denoted by $I(\mathbf{y})$ and $H(\mathbf{y})$, although they are defined for stochastic processes and not for their sample realizations.

time p. Thus, we achieve the joint block diagonalization of the output correlation matrices by minimizing $I(\mathbf{y})$ in (5.30). Using the fact that the determinant of the block-diagonal matrix is the product of the determinant of each diagonal block [25] and after replacing the true correlation matrix $\mathbf{R_{yy}}(p)$ by its estimate,[6] $\widehat{\mathbf{R}}_{\mathbf{yy}}(p)$ yields the following cost function:

$$J(\mathbf{W}) = \sum_{k=1}^{K} \log \det \operatorname{bdiag}(\widehat{\mathbf{R}}_{\mathbf{yy}}(kL)) - \log \det \widehat{\mathbf{R}}_{\mathbf{yy}}(kL), \qquad (5.34)$$

where $\operatorname{bdiag}(\mathbf{A}) = \mathbf{A} - \operatorname{boff}(\mathbf{A})$.

Graphical representation

Let us examine the cost function in (5.34) in a simple graphical way. Figure 5.2 depicts the negated cost function, $-J(\mathbf{W})$, whose representation is more readable than that of $J(\mathbf{W})$. The following setting is considered: The dimension of the instantaneous mixing ($L = L_{\mathrm{m}} = 1$) is set to $M = N = 2$. To fix the scaling ambiguity, the diagonal terms are set to $w_{nn,0} = 1$ for $n = 1, 2$. The varying parameters are the off-diagonal terms $w_{12,0}$ and $w_{21,0}$. $T = 100$ independent realizations of $\mathbf{s}$ are drawn from a Gaussian distribution with random variance. Correlation matrices $\mathbf{R_{xx}}(p) = \mathbf{E}\left\{\mathbf{x}(p)\mathbf{x}^{\mathrm{T}}(p)\right\}$ are estimated by the moving-average $\widehat{\mathbf{R}}_{\mathbf{xx}}(p) = \frac{1}{3}\sum_{\tau=-1}^{1} \mathbf{x}(p-\tau)\mathbf{x}^{\mathrm{T}}(p-\tau)$. The mixing matrix is set to

$$\mathbf{H} = \begin{pmatrix} 1 & -0.8 \\ -0.8 & 1 \end{pmatrix}.$$

It may be seen that J has only two local minima. They correspond to the two separating solutions, the "direct" one and the "permuted" one. These minima

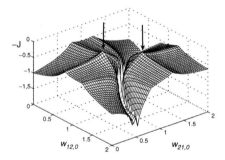

Fig. 5.2. Cost function J for an instantaneous mixing. For better readability, the negative J is shown. *Two arrows* indicate the minima of J (maxima of $-J$) corresponding to the separating solutions

---

[6] The implementation of $\widehat{\mathbf{R}}_{\mathbf{yy}}(p)$ is not specified for now. We refer to Sect. 6.1.5 for an implementation of second-order statistics BSS algorithms.

of J are separated by singularities of J (we have $J(\mathbf{W}) = +\infty$ for singular separation matrices $\mathbf{W}$).

5.3 Gradient-Based Minimization

5.3.1 Standard Gradient

A simple approach to find the separation matrix $\mathbf{W}$ that minimizes a cost function J is the iterative gradient descent, which may be written using a variable step-size $\mu(n)$ as follows:

$$\mathbf{W}(n + 1) = \mathbf{W}(n) - \mu(n) \left.\frac{\partial J}{\partial \mathbf{W}}\right|_{\mathbf{W}=\mathbf{W}(n)}. \tag{5.35}$$

A remark is in order regarding (5.35) in the case of Sylvester matrices. Sylvester matrices allow to formulate the block convolution very concisely, but face two related issues: Firstly, implementing (5.35) directly would involve updating $MNL(2L-1)$ terms instead of MNL free filter coefficients. Secondly, the gradient term $\frac{\partial J}{\partial \mathbf{W}}$ should have a Sylvester structure, which is not the case for the cost function in (5.34). These technical issues are ignored for now, because this chapter aims at introducing the main principles of BSS. They will be treated and discussed specifically in Chap. 6.

Since we have[7] $\mathbf{R_{yy}}(p) = \mathbf{WR_{xx}}(p)\mathbf{W}^{\mathrm{T}}$, the two formulas needed to obtain the gradient of (5.34) are

$$\frac{\partial \log \det \mathrm{bdiag}\mathbf{WR_{xx}}\mathbf{W}^{\mathrm{T}}}{\partial \mathbf{W}} = \mathrm{bdiag}^{-1}\left(\mathbf{WR_{xx}}\mathbf{W}^{\mathrm{T}}\right)\mathbf{WR_{xx}}, \tag{5.36}$$

$$\frac{\partial \log \det \mathbf{WR_{xx}}\mathbf{W}^{\mathrm{T}}}{\partial \mathbf{W}} = \left(\mathbf{WR_{xx}}\mathbf{W}^{\mathrm{T}}\right)^{-1}\mathbf{WR_{xx}}, \tag{5.37}$$

where $\mathrm{bdiag}^{-1}\mathbf{A} = (\mathrm{bdiag}\mathbf{A})^{-1}$ [20, 25].[8] Using (5.36) and (5.37), the gradient of the cost function (5.34) can be written as

$$\frac{\partial J}{\partial \mathbf{W}} = \sum_{k=1}^{K}\left(\mathrm{bdiag}^{-1}\left(\widehat{\mathbf{R}}_{\mathbf{yy}}(kL)\right) - \widehat{\mathbf{R}}_{\mathbf{yy}}^{-1}(kL)\right)\mathbf{W}\widehat{\mathbf{R}}_{\mathbf{xx}}(kL). \tag{5.38}$$

The inversion of the $NL \times NL$ matrices $\widehat{\mathbf{R}}_{\mathbf{yy}}^{-1}(kL)$ in (5.38) makes the standard gradient descent particularly unattractive.

---

[7] The arguments n and p may be omitted for notational convenience.

[8] The Sylvester structure of $\mathbf{W}$ is not explicitly taken into account in (5.36) and (5.37) (see [20, 25]). The resulting gradient has no Sylvester structure and maintaining the Sylvester structure requires a special treatment. This issue is treated in Chap. 6.

5.3.2 Natural Gradient

The natural gradient is an alternative to the standard gradient that applies especially well to the SOS-BSS cost function in (5.34). However, it is restricted to square scenarios where the number of sources equals the number of sensors ($N = M$). Derived by Amari from differential geometry considerations [7], it was independently introduced by Cardoso, who called it the "relative" gradient [23]. The natural gradient update may be obtained by modifying the standard gradient $\frac{\partial J}{\partial \mathbf{W}}$ as follows:

$$\Delta \mathbf{W} \triangleq \frac{\partial J}{\partial \mathbf{W}} \mathbf{W}^{\mathrm{T}} \mathbf{W}. \tag{5.39}$$

This formula was also derived for instantaneous mixtures. We will see in Chap. 6 how it may be extended to convolutive mixtures. Combining (5.38) and (5.39) yields

$$\Delta \mathbf{W} = \sum_{k=1}^{K} \mathrm{bdiag}^{-1} \widehat{\mathbf{R}}_{\mathbf{yy}}(kL) \mathrm{boff}\left(\widehat{\mathbf{R}}_{\mathbf{yy}}(kL)\right) \mathbf{W}. \tag{5.40}$$

The computation of the natural gradient learning terms in (5.40) is much less demanding than the computation of (5.38). Instead of inverting the $NL \times NL$ correlation matrix $\widehat{\mathbf{R}}_{\mathbf{yy}}(p)$, only the submatrices $\widehat{\mathbf{R}}_{\mathbf{y}_n \mathbf{y}_n}(p)$ of size $L \times L$ need to be inverted in (5.40). This is performed by inverting each diagonal block of size $L \times L$.

The algorithmic complexity of (5.40) may be further reduced by replacing $\mathrm{bdiag}^{-1} \widehat{\mathbf{R}}_{\mathbf{yy}}(p)$ with $\mathrm{diag}^{-1} \widehat{\mathbf{R}}_{\mathbf{yy}}(p)$, yielding

$$\Delta \mathbf{W} = \sum_{k=1}^{K} \mathrm{diag}^{-1} \widehat{\mathbf{R}}_{\mathbf{yy}}(kL) \mathrm{boff}\left(\widehat{\mathbf{R}}_{\mathbf{yy}}(kL)\right) \mathbf{W}. \tag{5.41}$$

Only scalar numbers need to be inverted in (5.41), as opposed to block-diagonal matrices as in (5.40). This simplification is obtained by approximating the correlation $\widehat{\mathbf{R}}_{\mathbf{y}_n \mathbf{y}_n}$ of $\mathbf{y}_n(p)$ by its power: $\mathbf{R}_{\mathbf{y}_n \mathbf{y}_n} \approx \widehat{\mathrm{E}}\{y_n^2(p)\}\mathbf{I}$. There is an analogy between algorithm (5.41) and the NLMS algorithm (4.7): In (5.41) and in (4.7), the normalization is performed by scalar power terms [20]. In the remainder of this work, the generic algorithm (5.41) is referred to as the natural gradient second-order statistics BSS algorithm (NG-SOS-BSS).

Nonholonomicity

If the output cross correlations vanish for all blocks, i.e., if $\mathrm{bdiag}\left(\mathbf{R}_{\mathbf{yy}}(kL)\right) = \mathbf{R}_{\mathbf{yy}}(kL)$ for $k = 1, \ldots, K$, then the update terms in (5.40) and in (5.41) vanish. This equilibrium condition sets no constraint on the power or on the self-correlation of the output signals. For this reason, these learning rules are referred to as *nonholonomic*.

Let us give a formal definition for the nonholonomicity in the context of BSS. Without loss of generality, any iterative algorithm for optimizing $\mathbf{W}$ may be written as

$$\mathbf{W}(n+1) = f\left(\mathbf{W}(n)\right). \tag{5.42}$$

The matrix function f depends not only on the input data but also on other parameters like the step-size. Let us consider a linear transform on the output signals $\mathbf{y}$, which is represented by the $NL \times NL$ block-diagonal matrix $\mathbf{D}$, and examine $\mathbf{Dy}$. This linear transform is equivalently expressed on the separation filters, since $\mathbf{Dy} = (\mathbf{DW})\mathbf{x}$. Hence, to simply transform the output data linearly, we define the function g as

$$g\left(\mathbf{W}, \mathbf{D}\right) \triangleq f\left(\mathbf{DW}\right) \tag{5.43}$$

for any $NL \times NL$ block-diagonal matrix $\mathbf{D}$. We can now rewrite (5.42) with the identity matrix $\mathbf{I}$:

$$\mathbf{W}(n+1) = g\left(\mathbf{W}(n), \mathbf{I}\right). \tag{5.44}$$

The equilibria of the algorithm in (5.42) are fixed points $\mathbf{W}^*$ of $g\left(\cdot, \mathbf{I}\right)$, that is, they are the points $\mathbf{W}^*$ so that

$$\mathbf{W}^* = g\left(\mathbf{W}^*, \mathbf{I}\right). \tag{5.45}$$

Now, the algorithm represented by (5.42) is said to be nonholonomic if and only if for all block-diagonal matrix $\mathbf{D}$, any fixed point $\mathbf{W}^*$ of $g\left(\cdot, \mathbf{I}\right)$ is also a fixed point of $g\left(\cdot, \mathbf{D}\right)$:

$$\mathbf{W}^* = g\left(\mathbf{W}^*, \mathbf{I}\right) \Rightarrow \mathbf{W}^* = g\left(\mathbf{W}^*, \mathbf{D}\right) \quad \forall \text{ diagonal matrix } \mathbf{D}. \tag{5.46}$$

The nonholonomicity is important if the source signals are speech signals because for holonomic learning rules, the equilibrium condition depends on the power or on the self-correlation of the output signals. Typically, holonomic learning rules tend to make the output signals temporally white and seem unsuitable in online processing of speech signals [8].

5.4 Natural Gradient Algorithm for NonSquare Systems

In the derivation of the natural gradient (5.39), it is assumed that there are as many sensors as sources [7, 23, 56]. Therefore, the learning rules (5.40) and (5.41) apply only if $M = N$. However, we may have more sensors than sources and exploiting the information provided by these additional sensors should improve the separation performance.

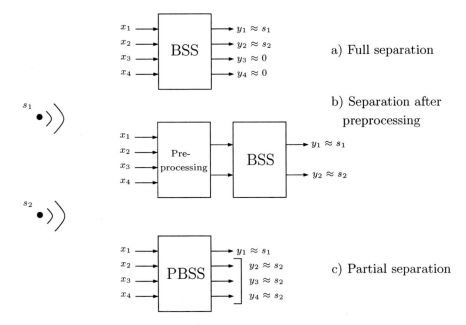

Fig. 5.3. Three approaches for extending the natural gradient to nonsquare systems

Full separation

A straightforward approach would be to overestimate the number of sources [93]. Setting $N = M$, the natural gradient may be applied for any number of sensors, as depicted in Fig. 5.3a. However, this is computationally demanding. Moreover, the information on the actual number of sources is not taken into account.

Separation after preprocessing

Another approach is to apply a fixed preprocessing to the microphone signals to reduce the dimension of the observed signals to N, as depicted in Fig. 5.3b. Using the principal component analysis (PCA) for example, one may select the components of the microphone signals that have the most energy [12, 56]. This approach yields the advantage of a lower complexity, since BSS is applied on $N < M$ input signals. However, it is not clear whether this preprocessing may discard information that is useful for separation or not.

Partial separation

The technique that we propose is a trade-off between the full separation and the separation after preprocessing and is referred to as partial BSS (PBSS). The idea is very simple: Since it is not necessary to extract M independent components, we may assign several outputs to *one* source, as depicted in Fig. 5.3c. In the case $N = 2$, $M = 4$, we may assign y_1 to the source s_1 and y_2, y_3, y_4 to the source s_2. We note that other assignments may also be possible. This particular assignment is well suited if the source s_1 is placed closest to the microphone x_1, since at the initialization we have $y_1 = x_1$. This particular assignment is also structurally similar to the spatial preprocessing performed in the GSC, which will reveal useful in Chap. 9.

Let us define the vector $\mathbf{y}_{[2,3,4]}(p)$ as

$$\mathbf{y}_{[2,3,4]}(p) \triangleq \left(\mathbf{y}_2^{\mathrm{T}}(p), \mathbf{y}_3^{\mathrm{T}}(p), \mathbf{y}_4^{\mathrm{T}}(p)\right)^{\mathrm{T}}. \tag{5.47}$$

The partial separation of the sources may be achieved by minimizing a cost function J_{PBSS} which is obtained by modifying the cost function (5.34). In Partial BSS, we do not want to minimize the mutual information $I(\mathbf{y}_1, \dots, \mathbf{y}_4)$. Here, the mutual information $I(\mathbf{y}_1, \mathbf{y}_{[2,3,4]})$ is relevant. In case of nonstationary Gaussian signals, this mutual information is given by

$$
\begin{aligned}
I\left(\mathbf{y}_1(p), \mathbf{y}_{[2,3,4]}(p)\right) &= \log \det \mathbf{R}_{\mathbf{y}_1 \mathbf{y}_1}(p) \\
&\quad + \log \det \mathbf{E}\left\{\mathbf{y}_{[2,3,4]}(p)\mathbf{y}_{[2,3,4]}^{\mathrm{T}}(p)\right\} \\
&\quad - \log \det \mathbf{R}_{\mathbf{y}\mathbf{y}}(p).
\end{aligned}
\tag{5.48}
$$

This yields the cost function

$$J_{\mathrm{PBSS}} = \sum_{k=1}^{K} \log \det \mathrm{bdiag}_{1,[2,3,4]}(\widehat{\mathbf{R}}_{\mathbf{y}\mathbf{y}}(kL)) - \log \det(\widehat{\mathbf{R}}_{\mathbf{y}\mathbf{y}}(kL)). \tag{5.49}$$

The matrix operator $\mathrm{bdiag}_{1,[2,3,4]}$ is defined as follows: Let $(\mathcal{K}_1, \mathcal{K}_2)$ be a partition of $\{1, \dots, M\}$, the operator $\mathrm{bdiag}_{\mathcal{K}_1, \mathcal{K}_2}(\mathbf{A})$ sets the (n,m)th blocks of the matrix $\mathbf{A}$ to zero for all $(n,m) \in (\mathcal{K}_1, \mathcal{K}_2)$ such that $n \neq m$. This simple operation may be made explicit in the case of $\mathrm{bdiag}_{1,[2,3,4]}$ as follows:

$$\mathrm{bdiag}_{1,[2,3,4]}
\begin{bmatrix}
\mathbf{A}_{11} & \cdots & \mathbf{A}_{14} \\
\vdots & \ddots & \vdots \\
\mathbf{A}_{41} & \cdots & \mathbf{A}_{44}
\end{bmatrix}
\triangleq
\begin{bmatrix}
\mathbf{A}_{11} & \mathbf{0} & \cdots & \mathbf{0} \\
\mathbf{0} & \mathbf{A}_{22} & \cdots & \mathbf{A}_{24} \\
\vdots & \vdots & \ddots & \vdots \\
\mathbf{0} & \mathbf{A}_{42} & \cdots & \mathbf{A}_{44}
\end{bmatrix}. \tag{5.50}$$

An existing approach that may be related to PBSS is the deflation approach [25]. It consists of two steps: (1) a single source is extracted and (2) removed from the mixture signals using a least-square power criterion. Both steps are repeated iteratively until a single source remains. PBSS is similar to the first step of the deflation approach. However, at the extraction of each source $\mathbf{y}_n(p)$, deflation algorithms typically exploit an optimization criterion of the form $J(\mathbf{y}_n)$ that depends on a single output; on the contrary, PBSS optimizes $\mathbf{W}$ for different outputs jointly with a criterion of the form $J(\mathbf{y}_1, \mathbf{y}_{[2,\dots,M]})$.

PBSS in the case $N = 2$

In the case of two sources $N = 2$, this approach may be generalized to M sensors replacing $\mathrm{bdiag}_{1,[2,3,4]}$ by $\mathrm{bdiag}_{1,[2,...,M]}$. In our context, we consider that only one source signal is actually desired. (For example in our car application, only the driver speech is actually desired.) The other outputs of the system are considered as by-product of the PBSS algorithm. In other words, the two sources actually consist in (1) the desired source and (2) the set of all other present sources. Hence, only the case of two sources $N = 2$ is further considered for PBSS implementations. However, we note that PBSS may also be generalized to $N > 2$ sources and we notice that PBSS coincides with BSS if $N = M$.

Similarly to the BSS cost function in (5.34), J_{PBSS} may be minimized with a natural gradient approach, which yields

$$\mathbf{\Delta W} = \sum_{k=1}^{K} \mathrm{bdiag}_{1,[2,...,M]}^{-1} \widehat{\mathbf{R}}_{\mathbf{yy}}(kL) \left(\widehat{\mathbf{R}}_{\mathbf{yy}}(kL) - \mathrm{bdiag}_{1,[2,...,M]} \widehat{\mathbf{R}}_{\mathbf{yy}}(kL) \right) \mathbf{W}.$$

$$(5.51)$$

The matrix inversion may be avoided using an NLMS-like algorithm as in (5.41). We then obtain the following NG-SOS-BSS learning rule:

$$\mathbf{\Delta W} = \sum_{k=1}^{K} \mathrm{diag}^{-1} \widehat{\mathbf{R}}_{\mathbf{yy}}(kL) \left(\widehat{\mathbf{R}}_{\mathbf{yy}}(kL) - \mathrm{bdiag}_{1,[2,...,M]} \widehat{\mathbf{R}}_{\mathbf{yy}}(kL) \right) \mathbf{W}. \quad (5.52)$$

The computational demand in PBSS may be further reduced if we do not adapt the cross filters $\mathbf{w}_{nm}, n \neq m$ relating outputs that are assigned to the same source $s_n(p)$. This should not impair the separation performance, since it is not necessary to reduce the cross correlation between these outputs. In the case $N = 2$, the filters $\mathbf{w}_{nm}$ for $n, m > 1$ and $n \neq m$ remain zero. This results in the separation matrix

$$\mathbf{W} = \begin{bmatrix} \mathbf{W}_{11} & \mathbf{W}_{12} & \cdots & \cdots & \mathbf{W}_{1M} \\ \mathbf{W}_{21} & \mathbf{W}_{22} & \mathbf{0} & \cdots & \mathbf{0} \\ \vdots & \mathbf{0} & \ddots & \ddots & \vdots \\ \vdots & \vdots & \ddots & \ddots & \mathbf{0} \\ \mathbf{W}_{M1} & \mathbf{0} & \cdots & \mathbf{0} & \mathbf{W}_{MM} \end{bmatrix}. \quad (5.53)$$

The PBSS approach realizes a trade-off between the full separation and the preprocessing approach depicted in Fig. 5.3a and b, respectively:

- It is computationally less demanding than full separation. The matrix

$$\widehat{\mathbf{R}}_{\mathbf{yy}}(kL) - \mathrm{bdiag}_{1,[2,...,M]}(\widehat{\mathbf{R}}_{\mathbf{yy}}(kL) \quad (5.54)$$

yields $2(M - 1)$ nonzero blocks while

$$\widehat{\mathbf{R}}_{\mathbf{yy}}(kL) - \text{bdiag}(\widehat{\mathbf{R}}_{\mathbf{yy}}(kL)) \tag{5.55}$$

yields $M(M-1)$ nonzero blocks. Moreover, if we do not adapt certain cross filters as in (5.53), the number of adaptive filters is reduced from M^2 to $3M - 2$.

- It offers more flexibility than the preprocessing approach. $M - 1$ interference signals are obtained as output instead of $N - 1$ for the preprocessing approach, which may be advantageous for postprocessing of the output signals. This feature will be exploited in Chap. 9. Also, since the M microphone signals are passed as inputs, PBSS may exploit the entire observed information.

5.5 Summary and Conclusion

This chapter introduced the main principles of Second-Order Statistics Blind Source Separation (SOS-BSS). BSS aims at identifying acoustic mixings by exploiting solely the mutual independence of the source signals. This can be achieved by exploiting the second-order statistics of nonstationary signals.

The formal framework for convolutive BSS has been introduced by representing filters with Sylvester matrices, as in [18]. Within this framework, a cost function that exploits the second-order statistics and nonstationarity of the source signals has been derived. The minimization of this cost function may be performed with the natural gradient. The computational demand of the resulting Natural Gradient SOS-BSSalgorithm (NG-SOS-BSS) is much lower than with the standard gradient descent. However, the natural gradient can be used with square systems only. Extensions of the natural gradient to nonsquare systems have been discussed.

Unfortunately, the gradient update (5.38) and the natural gradient (5.39) are valid only in the instantaneous case and the Sylvester structure of $\mathbf{W}$ in the convolutive case has been ignored. For this reason, the derived rules (5.38), (5.40), and (5.41) cannot be used directly. In Chap. 6, we will explain how the convolutive nature of the mixing may be taken into account.

6

Implementation Issues in Blind Source Separation

In Chap. 5, we have derived BSS algorithms using Sylvester matrices. The Sylvester-based notation allows the separation process to be written very concisely as in (5.14):

$$\mathbf{y}(p) = \mathbf{W}\mathbf{x}(p). \tag{5.14}$$

Paradoxically, the price for the conciseness of the Sylvester-based notation is its high redundancy. And unfortunately, consistency of this redundant representation of the separation system $\mathbf{W}$ is not guaranteed by the algorithms that are given in Chap. 5. For this reason, these algorithms cannot be directly implemented. In the first part of this chapter, our contribution is to propose a systematic and rigorous treatment of this redundancy issue, which eventually provides implementable update rules.

The second part of this chapter (Sects. 6.2 and 6.3) addresses other practical aspects of signal separation algorithms. Section 6.2 gives a general scheme for online implementations of BSS algorithms. Experimental results are presented in Sect. 6.3.

6.1 Natural Gradient in the Convolutive Case

Considering N sources, M microphones, and separation filters of length L, we need to adjust NML filter coefficients. However, the Sylvester matrix $\mathbf{W}$ in (5.11) has $NML(2L-1)$ entries. Ignoring the Sylvester structure leads to the gradient descent (5.38), where the gradient is calculated with respect to each entry of $\mathbf{W}$ independently. Therefore, the adaptation term $\mathbf{\Delta W} = \frac{\partial J}{\partial \mathbf{W}}$ has no Sylvester structure. The Sylvester structure of $\mathbf{\Delta W}$ may be restored by selecting NML nonredundant elements in $\mathbf{\Delta W}$. Then, one may build the matrix $\mathbf{\Delta W}$ in the Sylvester form (5.9) with these reference elements. For example, choosing the L first elements of the first row of $\mathbf{\Delta W}_{nm}$ (for each n, m) fulfills this nonredundancy requirement [18]. Alternatively, one may force the Sylvester structure by using the Lth column of $\mathbf{\Delta W}_{nm}$, as

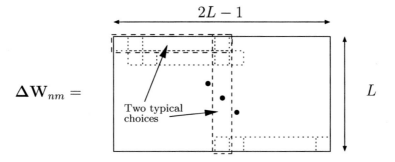

$$\mathbf{\Delta W}_{nm} =$$

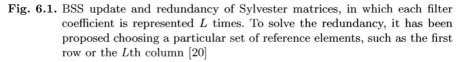

Fig. 6.1. BSS update and redundancy of Sylvester matrices, in which each filter coefficient is represented L times. To solve the redundancy, it has been proposed choosing a particular set of reference elements, such as the first row or the Lth column [20]

shown in Fig. 6.1 [20]. These two choices are related to certain convolution formulas, as observed by Aichner et al. [5]. At first glance, any other a priori *arbitrary* choice of nonredundant subsets of elements (or any combination of them) seem possible.

In this section, we tackle this problem and derive rigorously convolutive formulations of the natural gradient for a general cost function J. As a preliminary, the gradient $\frac{\partial J}{\partial \mathbf{W}}$ is derived for Sylvester matrices (Sect. 6.1.1). Secondly, we develop convolutive formulations of the natural gradient; for convenience, the derivation is carried out using z-transforms, but the resulting formulas are also expressed in the time domain (Sects. 6.1.2 to 6.1.4). Then, we apply these results to the BSS cost function in (5.34), and propose an approximation to obtain efficient update rules (Sect. 6.1.5). At last, their properties are discussed in Sect. 6.1.6.

6.1.1 Gradient in the Sylvester Subspace

If we consider the set of all $NL \times M(2L-1)$ matrices as an $NML(2L-1)$-dimensional vector space, the subset of the Sylvester matrices may be seen as an NML-dimensional subspace in this vector space. In the following, this subspace is called the Sylvester subspace[1] and is denoted by $\boldsymbol{S}$. This subsection explains how the gradient of a real function

$$f : \boldsymbol{S} \rightarrow \mathbb{R} \tag{6.1}$$

is computed in the Sylvester subspace.

Let us denote the (i,j)-entry of the matrix $\mathbf{W}_{nm}$ by w_{ij}^{nm} for $i = 1, \ldots, L$ and $j = 1, \ldots, 2L-1$. Expanding the argument $\mathbf{W}$, $f(\mathbf{W})$ may be written as

---

[1] By definition of the Sylvester matrices in (5.9), there is a one-to-one correspondence between the Sylvester subspace and the set of all FIR filters of length L, which is sometimes called "FIR manifold" [95].

$$f(\mathbf{W}) = f(w_{11}^{11}, \ldots, w_{ij}^{nm}, \ldots, w_{L,2L-1}^{NM}). \tag{6.2}$$

The function f is defined on the set of all $NL \times M(2L-1)$ matrices and its gradient is computed in all directions $\partial/\partial w_{ij}^{nm}$ independently. Let us consider a given filter coefficient $w_{nm,l}$. This coefficient appears L times in the matrix $\mathbf{W}$. We denote the indexes of the elements of $\mathbf{W}$ that are equal to $w_{nm,l}$ by $\{i_1 j_1, \ldots, i_l j_l, \ldots, i_L j_L\}$ and the restriction of f to the parameter $w_{nm,l}$ by $f|_{nm,l}$. To obtain the gradient with respect to $w_{nm,l}$, we consider a small deviation ε around $w_{nm,l}$, i.e., around $w_{i_l j_l}^{nm}$ for all $l = 1, \ldots, L$ and write the first-order development of the function $f|_{nm,l}$:

$$f|_{nm,l}(w_{nm,l} + \varepsilon) \triangleq f(\ldots, w_{nm,l} + \varepsilon, \ldots, w_{nm,l} + \varepsilon, \ldots), \tag{6.3}$$

$$= f(\ldots, w_{nm,l}, \ldots) + \sum_{k=1}^{L} \varepsilon \frac{\partial f}{\partial w_{i_k j_k}^{nm}}(w_{nm,l}) + o(\varepsilon^2), \tag{6.4}$$

$$= f|_{nm,l}(w_{nm,l}) + \varepsilon \sum_{k=1}^{L} \frac{\partial f}{\partial w_{i_k j_k}^{nm}}(w_{nm,l}) + o(\varepsilon^2). \tag{6.5}$$

By definition, the derivative $\partial f|_{nm,l}/\partial w_{nm,l}$ is so that

$$f|_{nm,l}(w_{nm,l} + \varepsilon) = f|_{nm,l}(w_{nm,l}) + \varepsilon \frac{\partial f|_{nm,l}}{\partial w_{nm,l}}(w_{nm,l}) + o(\varepsilon^2). \tag{6.6}$$

Combining the definition (6.6) and (6.5) directly yields

$$\frac{\partial f|_{nm,l}}{\partial w_{nm,l}}(w_{nm,l}) = \sum_{k=1}^{L} \frac{\partial f}{\partial w_{i_k j_k}^{nm}}(w_{nm,l}). \tag{6.7}$$

The restriction of J to the Sylvester subspace $\mathbf{S}$ is denoted by $J|_S$. Equation (6.7) tells us that the gradient $\frac{\partial J|_S}{\partial \mathbf{W}}$ of the cost function $J|_S$ is simply generated in the form of (5.9) from the *sums* of all redundant terms of $\frac{\partial J}{\partial \mathbf{W}}$.

Approximations of $\frac{\partial J|_S}{\partial \mathbf{W}}$

The computation of $\frac{\partial J|_S}{\partial \mathbf{W}}$ is expensive: At each step, the whole matrix $\frac{\partial J}{\partial \mathbf{W}}$ needs to be computed and made Sylvester by summing. For later reference, we denote by S the operator that transforms a general matrix into a block Sylvester matrix, i.e., $\frac{\partial J|_S}{\partial \mathbf{W}} = S(\frac{\partial J}{\partial \mathbf{W}})$. This operator is formally defined in (6.9). The choice of a particular subset as reference to impose the Sylvester constraint may be seen as an approximation of S. If this reference is taken at the dth row, the approximation of S is denoted by S_d and is defined in (6.8) for a one-block matrix $\mathbf{A} = [a]_{ij}$. If the reference is taken at the Lth column, the approximation of S is denoted by S^L and is defined in (6.10) for a one-block matrix $\mathbf{A} = [a]_{ij}$. The matrix operators S_d, S, and S^L are defined in (6.8), (6.9), and (6.10) for a one-block $L \times 2L - 1$ matrix $\mathbf{A}$ with coefficients a_{ij}:

$$S_d(\mathbf{A}) \triangleq \begin{pmatrix} a_{dd} & a_{d(d+1)} & \cdots & a_{d(d+L-1)} & 0 & \cdots & 0 \\ 0 & \ddots & & \vdots & & \ddots & \vdots \\ \vdots & & \ddots & \vdots & & & 0 \\ 0 & \cdots & 0 & a_{dd} & & \cdots & a_{d(d+L-1)} \end{pmatrix}, \tag{6.8}$$

$$S(\mathbf{A}) \triangleq \sum_{d=1}^{L} S_d(\mathbf{A}), \tag{6.9}$$

$$S^L(\mathbf{A}) \triangleq \begin{pmatrix} a_{LL} & a_{(L-1)L} & \cdots & a_{1L} & 0 & \cdots & 0 \\ 0 & \ddots & & \vdots & \ddots & \ddots & \vdots \\ \vdots & & \ddots & & \vdots & & 0 \\ 0 & \cdots & & 0 & a_{LL} & \cdots & a_{1L} \end{pmatrix}. \tag{6.10}$$

Proportionality principle

Without loss of generality, we can write the nth iteration of the minimization as

$$\mathbf{W}(n+1) = \mathbf{W}(n) + \Delta\mathbf{W}(n), \tag{6.11}$$

where $\mathbf{W}(n), \Delta\mathbf{W}(n) \in S$. The natural gradient requires that the learning term $\Delta\mathbf{W}(n)$ is "proportional" to the current separation matrix $\mathbf{W}(n)$ in the sense that $\Delta\mathbf{W}(n)$ can be written as

$$\Delta\mathbf{W}(n) = \mathbf{D}(n) \star_S \mathbf{W}(n) \tag{6.12}$$

for a certain matrix $\mathbf{D}(n)$ and for a product $\star_S$ to be defined (see for example [22, 56]). To guarantee the consistency of the redundant Sylvester matrices, this product must maintain $\Delta\mathbf{W}(n)$ inside of the set S. Unfortunately, this requirement is too restrictive if we work with the standard matrix product $\mathbf{A} \star_S \mathbf{B} = \mathbf{AB}$: For $\mathbf{W}(n) \in S$, $\Delta\mathbf{W}(n) = \mathbf{D}(n)\mathbf{W}(n)$ belongs to S if and only if each submatrix $\mathbf{D}_{ij}(n)$ of $\mathbf{D}(n)$ is of the form $\mathbf{D}_{ij}(n) = d_{ij}\mathbf{I}$ for a scalar $d_{ij} \in \mathbb{R}$. (This statement may be easily verified.) Therefore, we leave the Sylvester matrices for a moment and work with the z-transform representation of the separation system.

6.1.2 From Matrices to z-Transforms

In this subsection, we introduce the notations and operators that are used to manipulate z-transforms. We assume that the number of source signals equals the number of microphones ($N = M$).

z-transform spaces

In the following, single matrix subscripts denote a time index (for example k in $\mathbf{A}_k$), while two subscripts denote a channel index (for example n, m in $\mathbf{A}_{nm}$). An element $\mathbf{A} \in \mathbf{S}$ is a collection of $N \times N$ filters of length L, with filter weights $a_{nm,k}, k = 0 \ldots, L-1$. Therefore, this element $\mathbf{A}$ is conveniently represented with its multichannel z-transform, which is denoted by $\overline{\mathbf{A}}$ and is defined as

$$\overline{\mathbf{A}}(z) \triangleq \sum_{k=0}^{L-1} \mathbf{A}_k z^{-k}, \tag{6.13}$$

$$\text{with } \mathbf{A}_k \triangleq \begin{pmatrix} a_{11,k} & \cdots & a_{1M,k} \\ \vdots & \ddots & \vdots \\ a_{N1,k} & \cdots & a_{NN,k} \end{pmatrix} \in \mathbb{R}^{N \times N} \tag{6.14}$$

for $k = 0, \ldots, L-1$. We denote the space of the multichannel *single-sided* z-transforms by $\overline{\mathbf{S}}$. In addition, we introduce the space $\overline{\mathbf{T}}$ of the multichannel *two-sided* z-transforms, which can be written as[2] $\overline{\mathbf{A}}(z) = \sum_{k=-\infty}^{\infty} \mathbf{A}_k z^{-k}$. If the support[3] of $\overline{\mathbf{A}}$ is included in $[-L+1, L-1]$, then $\overline{\mathbf{A}}$ may be represented as a matrix $\mathbf{A}$ whose submatrices $\mathbf{A}_{nm}$ have the Toeplitz structure:

$$\mathbf{A} = \begin{bmatrix} \mathbf{A}_{11} & \cdots & \mathbf{A}_{1N} \\ \vdots & \ddots & \vdots \\ \mathbf{A}_{N1} & \cdots & \mathbf{A}_{NN} \end{bmatrix} \tag{6.15}$$

$$\text{with } \mathbf{A}_{nm} = \begin{pmatrix} a_{nm,0} & a_{nm,1} & \cdots & a_{nm,L-1} \\ a_{nm,-1} & a_{nm,0} & \cdots & a_{nm,L-2} \\ \vdots & & \ddots & \ddots & \vdots \\ a_{nm,-L+1} & a_{nm,-L+2} & \cdots & a_{nm,0} \end{pmatrix} \tag{6.16}$$

for $n, m = 1, \ldots, N$. The set of the block Toeplitz matrices of size $NL \times NL$ is denoted by $\mathbf{T}$.

Linear operators

It is useful to introduce the operator $[\cdot]_{\mathcal{S}}$ that truncates a z-transform to a support $\mathcal{S} \subset [-L+1, L-1]$:

$$\left[\overline{\mathbf{B}}\right]_{\mathcal{S}}(z) \triangleq \sum_{k \in \mathcal{S}} \mathbf{B}_k z^{-k} \quad \forall z \in \mathbb{C}^*. \tag{6.17}$$

---

[2] Two-sided doubly infinite z-transforms are introduced for convenience but in fact, only finite sums are involved in the following. Therefore, $\overline{\mathbf{A}}(z)$ exists for any nonzero complex z.

[3] The support of $\overline{\mathbf{A}}$ is the set of integers k for which $\mathbf{A}_k \neq \mathbf{0}$. Again, note that $\mathbf{A}_k$ (with a single subscript k) is different from $\mathbf{A}_{nm}$ (with two subscripts n, m). The former represents the time index while the latter represents the channel index.

$\mathbb{C}^*$ denotes the set of the nonzero complex numbers, i.e., $\mathbb{C}^* = \mathbb{C}\backslash\{0\}$. $\left[\overline{\mathbf{B}}\right]_{\mathcal{S}}(z)$ is incorrectly[4] but commonly denoted by $\left[\overline{\mathbf{B}}(z)\right]_{\mathcal{S}}$. Any linear matrix operator f, for example $f(\cdot) = \mathrm{tr}(\cdot)$, may be defined on $\overline{T}$ using the following definition:

$$f(\overline{\mathbf{A}})(z) \triangleq \sum_k f\left(\mathbf{A}_k z^{-k}\right) \quad \forall \overline{\mathbf{A}} \in \overline{T}. \tag{6.18}$$

According to (6.18), the Hermitian conjugate of $\overline{\mathbf{A}} \in \overline{T}$, which is denoted by $\overline{\mathbf{A}}^H$, is defined as

$$\overline{\mathbf{A}}^H(z) \triangleq \sum_k \mathbf{A}_k^{\mathrm{T}} z^k. \tag{6.19}$$

The binary operator $\star_d$

We define the binary operator $\star_d$ as

$$\left(\overline{\mathbf{A}} \star_d \overline{\mathbf{B}}\right)(z) \triangleq \left[z^d \overline{\mathbf{A}}(z)\overline{\mathbf{B}}(z)\right]_{[0,L-1]} \tag{6.20}$$

for an integer $d \in [0, L-1]$. This definition is not arbitrary. Consider for example the case $d = 0$:

$$\left(\overline{\mathbf{A}} \star_0 \overline{\mathbf{B}}\right)(z) = \left[\overline{\mathbf{A}}(z)\overline{\mathbf{B}}(z)\right]_{[0,L-1]}. \tag{6.21}$$

The operation $\overline{\mathbf{A}} \star_0 \overline{\mathbf{B}}$ truncates the convolution $\overline{\mathbf{A}}\overline{\mathbf{B}}$, which has length $2L-1$ in the time domain, such that any terms of order higher than $L-1$ are omitted. The truncation assures that $\star_d$ is an internal operation in $\overline{S}$. The integer d allows for manipulation of acausal filters by varying the acausal length in the convolution. This becomes more obvious when we reformulate $\star_d$ as[5]

$$\left(\overline{\mathbf{A}} \star_d \overline{\mathbf{B}}\right)(z) = z^{-d}\left[z^d \overline{\mathbf{A}}(z)z^d \overline{\mathbf{B}}(z)\right]_{[-d,L-1-d]}. \tag{6.22}$$

Let us describe the operations that are realized in (6.22): $z^d \overline{\mathbf{A}}(z)$ shifts $\overline{\mathbf{A}}(z)$ so that the first d terms are acausal. The product $z^d \overline{\mathbf{A}}(z)z^d \overline{\mathbf{B}}(z)$ realizes the convolution where the d first coefficients of $\overline{\mathbf{A}}(z)$ and $\overline{\mathbf{B}}(z)$ are treated as acausal. The result is truncated on L coefficients and the term z^{-d} shifts the results back so that $\star_d$ is an internal operation in $\overline{S}$.

Scalar product for S and $\overline{T}$

Finally, we provide S with the scalar product $\langle \cdot, \cdot \rangle$ associated to the Euclidean metric:

$$\langle \mathbf{A}, \mathbf{B} \rangle \triangleq \mathrm{tr}\left(\mathbf{A}^{\mathrm{T}}\mathbf{B}\right) \quad \forall \mathbf{A}, \mathbf{B} \in S. \tag{6.23}$$

---

[4] Strictly speaking, for a given z, $\overline{\mathbf{B}}(z)$ is a complex matrix, not a polynomial.
[5] (6.22) is derived from the formula $z^{-d}[\overline{A}(z)]_{[p,q]} = [z^{-d}\overline{A}(z)]_{[p+d,q+d]}$.

Examination of $\langle \cdot, \cdot \rangle$ with Sylvester matrices reveals that

$$\forall \mathbf{A}, \mathbf{B} \in S \quad \langle \mathbf{A}, \mathbf{B} \rangle = L\mathrm{tr}\left(\sum_{k=0}^{L-1} \overline{\mathbf{A}}_k^{\mathrm{T}} \overline{\mathbf{B}}_k \right) = L\mathrm{tr}\left(\left[\overline{\mathbf{A}}^H \overline{\mathbf{B}} \right]_0 \right) = L \left[\mathrm{tr}\left(\overline{\mathbf{A}}^H \overline{\mathbf{B}} \right) \right]_0,$$

(6.24)

where the operator $[\overline{\mathbf{A}}]_0$ returns the constant coefficient of $\overline{\mathbf{A}}(z)$. We define the scalar product on $\overline{T}$ as

$$\langle \overline{\mathbf{A}}, \overline{\mathbf{B}} \rangle \triangleq L \left[\mathrm{tr}\left(\overline{\mathbf{A}}^H \overline{\mathbf{B}} \right) \right]_0$$

(6.25)

with the immediate property that

$$\forall \overline{\mathbf{A}}, \overline{\mathbf{B}} \in \overline{S} \quad \langle \overline{\mathbf{A}}, \overline{\mathbf{B}} \rangle = \langle \mathbf{A}, \mathbf{B} \rangle.$$

(6.26)

We now have the necessary tools to derive the natural gradient rigorously.

6.1.3 Self-Closed and Non-Self-Closed Natural Gradients

This subsection derives expressions of the natural gradient in terms of z-transforms.

Self-closed natural gradient

Let us denote the z-transform of the adaptation term $\mathbf{\Delta W}(n)$ in (6.11) by[6] $\overline{\mathbf{\Delta W}}$. We require that $\overline{\mathbf{\Delta W}}$ is "proportional" to $\overline{\mathbf{W}}$, in the sense that

$$\exists \overline{\mathbf{D}} \in \overline{S} \quad \overline{\mathbf{\Delta W}} = \overline{\mathbf{D}} \star_d \overline{\mathbf{W}}.$$

(6.27)

Note that the proportionality term $\overline{\mathbf{D}}$ is taken from the set $\overline{S}$, not from $\overline{T}$. Let us consider the derivations in (6.5) again: The first-order development of $J(\mathbf{W}(n+1))$ around $\mathbf{W}(n)$ with a small deviation $\mathbf{\Delta W}(n) \in S$ is given by

$$J(\mathbf{W}(n+1)) = J(\mathbf{W}(n)) + \left\langle \frac{\partial J}{\partial \mathbf{W}}, \mathbf{\Delta W}(n) \right\rangle + o(\|\mathbf{\Delta W}(n)\|^2).$$

(6.28)

According to the property in (6.26), we can reformulate (6.28) using $\overline{\mathbf{\Delta W}}$ as follows:

$$J(\mathbf{W}(n+1)) = J(\mathbf{W}(n)) + \left\langle \overline{\left(\frac{\partial J}{\partial \mathbf{W}} \right)}, \overline{\mathbf{\Delta W}} \right\rangle + o(\|\overline{\mathbf{\Delta W}}\|^2),$$

(6.29)

where $\overline{\left(\frac{\partial J}{\partial \mathbf{W}} \right)} \in \overline{S}$ is the z-transform of the gradient $\frac{\partial J|_s}{\partial \mathbf{W}}$. According to (6.29), the decrement $J(\mathbf{W}(n+1)) - J(\mathbf{W}(n))$ may be written in the first order as $\left\langle \overline{\left(\frac{\partial J}{\partial \mathbf{W}} \right)}, \overline{\mathbf{\Delta W}} \right\rangle$. Now, $\left\langle \overline{\left(\frac{\partial J}{\partial \mathbf{W}} \right)}, \overline{\mathbf{\Delta W}} \right\rangle$ can be rewritten using (6.24) as follows:

---

[6] The iteration index n is dropped for the sake of brevity.

$$\left\langle \overline{\left(\frac{\partial J}{\partial \mathbf{W}}\right)}, \overline{\Delta \mathbf{W}} \right\rangle = L \text{tr} \left(\left[\overline{\left(\frac{\partial J}{\partial \mathbf{W}}\right)}^H \overline{\mathbf{D} \star_d \mathbf{W}} \right]_0 \right), \tag{6.30}$$

$$= L \text{tr} \left(\left[\overline{\left(\frac{\partial J}{\partial \mathbf{W}}\right)}^H [z^d \overline{\mathbf{D} \mathbf{W}}]_{[0,L-1]} \right]_0 \right), \tag{6.31}$$

$$= L \text{tr} \left(\left[\overline{\left(\frac{\partial J}{\partial \mathbf{W}}\right)}^H z^d \overline{\mathbf{D} \mathbf{W}} \right]_0 \right). \tag{6.32}$$

From (6.31) and (6.32), we have used the fact that the support of $\overline{\left(\frac{\partial J}{\partial \mathbf{W}}\right)}$ is $[0, L-1]$. Therefore, we have $\left\langle \overline{\left(\frac{\partial J}{\partial \mathbf{W}}\right)}, [\overline{\mathbf{A}}]_{[0,L-1]} \right\rangle = \left\langle \overline{\left(\frac{\partial J}{\partial \mathbf{W}}\right)}, \overline{\mathbf{A}} \right\rangle$ for all $\overline{\mathbf{A}} \in \overline{T}$. Next, we recall that $\text{tr}\,(\overline{\mathbf{A} \mathbf{B}}) = \text{tr}\,(\overline{\mathbf{B} \mathbf{A}})$ for all $\overline{\mathbf{A}}, \overline{\mathbf{B}} \in \overline{T}$. Applying this property to (6.32) yields

$$\left\langle \overline{\left(\frac{\partial J}{\partial \mathbf{W}}\right)}, \overline{\Delta \mathbf{W}} \right\rangle = L \text{tr} \left(\left[\left(z^{-d} \overline{\left(\frac{\partial J}{\partial \mathbf{W}}\right)} \overline{\mathbf{W}}^H \right)^H \overline{\mathbf{D}} \right]_0 \right), \tag{6.33}$$

$$= \left\langle z^{-d} \overline{\left(\frac{\partial J}{\partial \mathbf{W}}\right)} \overline{\mathbf{W}}^H, \overline{\mathbf{D}} \right\rangle. \tag{6.34}$$

Now we use the fact that $\overline{\mathbf{D}}$ has support $[0, L-1]$, and thus we can reformulate (6.34) as

$$\left\langle \overline{\left(\frac{\partial J}{\partial \mathbf{W}}\right)}, \overline{\Delta \mathbf{W}} \right\rangle = \left\langle \left[z^{-d} \overline{\left(\frac{\partial J}{\partial \mathbf{W}}\right)} \overline{\mathbf{W}}^H \right]_{[0,L-1]}, \overline{\mathbf{D}} \right\rangle. \tag{6.35}$$

Let us choose $\overline{\mathbf{D}}$ such that the decrement $J(\mathbf{W}(n)) - J(\mathbf{W}(n+1))$ is maximized. According to (6.35), $\overline{\mathbf{D}}$ must be proportional to $-\left[z^{-d} \overline{\left(\frac{\partial J}{\partial \mathbf{W}}\right)} \overline{\mathbf{W}}^H \right]_{[0,L-1]}$, that is,

$$\overline{\mathbf{D}} = -\mu \left[z^{-d} \overline{\left(\frac{\partial J}{\partial \mathbf{W}}\right)} \overline{\mathbf{W}}^H \right]_{[0,L-1]} \tag{6.36}$$

for some $\mu > 0$. Substituting $\overline{\mathbf{D}}$ from (6.36) into (6.27) yields

$$\overline{\Delta \mathbf{W}} = -\mu \left[\left[\overline{\left(\frac{\partial J}{\partial \mathbf{W}}\right)} \overline{\mathbf{W}}^H \right]_{[-d, L-1-d]} \overline{\mathbf{W}} \right]_{[0,L-1]}. \tag{6.37}$$

Since $\overline{\mathbf{D}}$ is chosen in the same set $\overline{S}$ as the separation matrix $\overline{\mathbf{W}}$, the update rule in (6.37) is said to be *self-closed* [95]. The self-closed update (6.37) depends on the delay parameter d. The choice of this parameter will be discussed in Sect. 6.1.6.

Non-self-closed natural gradient

Let us weaken the requirement on $\overline{\mathbf{D}}$ in (6.27): We now choose $\overline{\mathbf{D}}$ from the set $\overline{\mathbf{T}}$. In this case, we can see from (6.34) that $\overline{\mathbf{D}}$ must be proportional to $z^{-d}\left(\overline{\frac{\partial J}{\partial \mathbf{W}}}\right)\overline{\mathbf{W}}^H$. Substituting this optimal $\overline{\mathbf{D}}$ into (6.27) yields

$$\overline{\Delta \mathbf{W}} = -\mu \left[\left(\overline{\frac{\partial J}{\partial \mathbf{W}}}\right)\overline{\mathbf{W}}^H \overline{\mathbf{W}}\right]_{[0,L-1]}. \tag{6.38}$$

We note that $\left(\overline{\frac{\partial J}{\partial \mathbf{W}}}\right)\overline{\mathbf{W}}^H$ has support $[-L+1, L-1]$ and may be represented as a Toeplitz matrix. Moreover, the dependence on d vanishes in (6.38). Since $\overline{\mathbf{D}}$ is chosen from the set $\overline{\mathbf{T}}$, which is a superset of $\overline{\mathbf{S}}$, the update rule in (6.38) is not self-closed.

6.1.4 From z-Transforms Back to the Time Domain

Now, the z-transform representation of the learning rules in (6.37) and (6.38) raises the question of how these learning rules can be formulated in the time domain in terms of Sylvester matrices. This section answers that question.

First, we consider the product $\left(\overline{\frac{\partial J}{\partial \mathbf{W}}}\right)\overline{\mathbf{W}}^H$ whose computation appears in both (6.37) and (6.38). The expression of this product in the time domain is explained in the following remark:

(*i*) For any $(\overline{\mathbf{G}}, \overline{\mathbf{W}}) \in \overline{\mathbf{S}} \times \overline{\mathbf{S}}$, the coefficients of $\overline{\mathbf{A}} = \overline{\mathbf{G}}\overline{\mathbf{W}}^H \in \overline{\mathbf{T}}$ are in the first row and first column of all submatrices of $\mathbf{A} = \mathbf{G}\mathbf{W}^T$, as illustrated in (6.39) in the case $L = 3$ and $M = N = 1$. These submatrices have a Toeplitz structure as in (6.16) and we have $\mathbf{G}\mathbf{W}^T \in \mathbf{T}$.

$$
\begin{array}{ccccc}
\overline{\mathbf{A}} & = & \overline{\mathbf{G}} & & \overline{\mathbf{W}}^H \\
\end{array}
$$

$$
\begin{pmatrix} a_0 & a_1 & a_2 \\ a_{-1} & a_0 & a_1 \\ a_{-2} & a_{-1} & a_0 \end{pmatrix} = \begin{pmatrix} g_0 & g_1 & g_2 & 0 & 0 \\ 0 & g_0 & g_1 & g_2 & 0 \\ 0 & 0 & g_0 & g_1 & g_2 \end{pmatrix} \begin{pmatrix} w_0 & 0 & 0 \\ w_1 & w_0 & 0 \\ w_2 & w_1 & w_0 \\ 0 & w_2 & w_1 \\ 0 & 0 & w_2 \end{pmatrix} \tag{6.39}
$$

The matrix $\mathbf{G}$ represents the gradient $\frac{\partial J}{\partial \mathbf{W}}$ and the matrix $\mathbf{A}$ represents the product $\frac{\partial J}{\partial \mathbf{W}}\mathbf{W}^T$.

Second, the computation of (6.37) and (6.38) involves postmultiplying by $\overline{\mathbf{W}}$ and truncating the result. Postmultiplication and truncation are explained in remark (*ii*) for (6.38) and in remark (*iii*) for (6.37).

(*ii*) For any $(\overline{\mathbf{A}}, \overline{\mathbf{W}}) \in \overline{\mathbf{T}} \times \overline{\mathbf{S}}$ such that the support of $\overline{\mathbf{A}}$ is included in $[-L+1, L-1]$, the coefficients of

$$\overline{\mathbf{B}} = [\overline{\mathbf{A}}\,\overline{\mathbf{W}}]_{[0,L-1]}$$

are in the Lth column of all submatrices of $\mathbf{B} = \mathbf{AW}$. This is illustrated in (6.40), where the symbol $\diamond$ represents a nonrelevant matrix entry:

$$
\begin{array}{ccc}
\overline{\mathbf{B}} & = & [\,\overline{\mathbf{A}} \qquad\qquad \overline{\mathbf{W}}\,]_{[0,L-1]} \\
\begin{pmatrix} \diamond\diamond\, b_2\, \diamond\diamond \\ \diamond\diamond\, b_1\, \diamond\diamond \\ \diamond\diamond\, b_0\, \diamond\diamond \end{pmatrix} & = & \begin{pmatrix} a_0 & a_1 & a_2 \\ a_{-1} & a_0 & a_1 \\ a_{-2} & a_{-1} & a_0 \end{pmatrix} \begin{pmatrix} w_0 & w_1 & w_2 & 0 & 0 \\ 0 & w_0 & w_1 & w_2 & 0 \\ 0 & 0 & w_0 & w_1 & w_2 \end{pmatrix}.
\end{array}
\tag{6.40}
$$

Note that $\mathbf{B}$ has no Sylvester structure.

(iii) For any $(\overline{\mathbf{A}}, \overline{\mathbf{W}}) \in \overline{\boldsymbol{T}} \times \overline{\boldsymbol{S}}$ and for $d \in \{0, \ldots, L\}$, the coefficients of

$$
\overline{\mathbf{D}} = [[\overline{\mathbf{A}}]_{[-d, L-1-d]} \overline{\mathbf{W}}]_{[0, L-1]}
$$

are in the $(d+1)$th row of all submatrix of $\mathbf{D} = \mathbf{AW}$, starting at the $(d+1)$th column. This is illustrated in (6.41) in the case $d = 1$:

$$
\begin{array}{ccc}
\overline{\mathbf{D}} & = & [\,[\overline{\mathbf{A}}]_{[-d, L-1-d]} \qquad\qquad \overline{\mathbf{W}}\,]_{[0,L-1]} \\
\begin{pmatrix} \diamond & \diamond & \diamond & \diamond & \diamond \\ \diamond & d_0 & d_1 & d_2 & \diamond \\ \diamond & \diamond & \diamond & \diamond & \diamond \end{pmatrix} & = & \begin{pmatrix} a_0 & a_1 & a_2 \\ 0 & a_0 & a_1 \\ 0 & 0 & a_0 \end{pmatrix} \begin{pmatrix} w_0 & w_1 & w_2 & 0 & 0 \\ 0 & w_0 & w_1 & w_2 & 0 \\ 0 & 0 & w_0 & w_1 & w_2 \end{pmatrix}.
\end{array}
\tag{6.41}
$$

The matrix $\mathbf{B}$ (resp. $\mathbf{D}$) represents the natural gradient $\frac{\partial J}{\partial \mathbf{W}} \mathbf{W}^{\mathrm{T}} \mathbf{W}$ in the non-self-closed case (resp. self-closed). The natural gradient is defined relative to the set $\overline{\boldsymbol{S}}$ or $\overline{\boldsymbol{T}}$ where $\overline{\mathbf{D}}$ is chosen. Combining remarks (i), (ii) and (iii) from above, we can find the natural gradient adaptation weights in the matrix $S(\frac{\partial J}{\partial \mathbf{W}}) \mathbf{W}^{\mathrm{T}} \mathbf{W}$.

- First, consider the self-closed case $\overline{\mathbf{D}} \in \overline{\boldsymbol{S}}$. There, the natural gradient in (6.37) additionally depends on $0 \leq d < L$. The natural gradient weights are obtained in the $(d+1)$th row of $S(\frac{\partial J}{\partial \mathbf{W}}) \mathbf{W}^{\mathrm{T}} \mathbf{W}$ as shown in (6.41) for $d = 1$. This is expressed using S and S_d (defined in (6.8)) as follows:

$$
\Delta \mathbf{W} = -\mu S_{d+1} \left(S \left(\frac{\partial J}{\partial \mathbf{W}} \right) \mathbf{W}^{\mathrm{T}} \mathbf{W} \right).
\tag{6.42}
$$

- Second, consider the case of the non-self-closed natural gradient, that is $\overline{\mathbf{D}} \in \overline{\boldsymbol{T}}$. The filter coefficients in (6.38) are obtained in the Lth column of $S(\frac{\partial J}{\partial \mathbf{W}}) \mathbf{W}^{\mathrm{T}} \mathbf{W}$ as shown in (6.40). Therefore, consistency of the update may be maintained using the operators S and S^L defined in (6.9) and (6.10) as follows:

$$
\Delta \mathbf{W} = -\mu S^L \left(S \left(\frac{\partial J}{\partial \mathbf{W}} \right) \mathbf{W}^{\mathrm{T}} \mathbf{W} \right).
\tag{6.43}
$$

Now, we can use the results of Sect. 5.3.2, where an expression of $\frac{\partial J}{\partial \mathbf{W}} \times \mathbf{W}^{\mathrm{T}}(n) \mathbf{W}(n)$ has been derived.

6.1.5 Application to NG-SOS-BSS

Approximating the updates in (6.42) and (6.43)

The detour around the z-transform shows how the Sylvester constraint should be implemented. Unfortunately, the updates we have obtained in (6.42) and (6.43) involve the computation of the Sylvester-space gradient $S(\frac{\partial J}{\partial \mathbf{W}})$, which is very demanding since the entire $NL \times M(2L-1)$ matrix $\frac{\partial J}{\partial \mathbf{W}}$ needs to be computed and made Sylvester by summing, as mentioned in Sect. 6.1.1. Moreover, in the case of the Gaussian mutual-information cost function (5.34), the gradient matrix $\frac{\partial J}{\partial \mathbf{W}}$ involves the inversion of output correlation matrices $\mathbf{R_{yy}}(kL)$ which may be badly conditioned especially for large filter length L and colored input signals (such as speech). To remove this matrix inversion and to benefit from the advantages of the natural gradient explained in Sect. 5.3.2, we need to alter the updates in (6.42) and (6.43).

For the self-closed update, if we remove $S()$, then (6.42) becomes

$$\mathbf{\Delta W} = -\mu S_{d+1}\left(\frac{\partial J}{\partial \mathbf{W}}\mathbf{W}^\mathrm{T}\mathbf{W}\right). \tag{6.44}$$

In the case of the non-self-closed natural gradient, if we remove $S()$, then (6.43) becomes

$$\mathbf{\Delta W} = -\mu S^L\left(\frac{\partial J}{\partial \mathbf{W}}\mathbf{W}^\mathrm{T}\mathbf{W}\right). \tag{6.45}$$

In both cases, the natural gradient updates (6.44) and (6.45) may be summarized using an approximation S_{approx} of S as

$$\Delta\mathbf{W}(n) = -\mu(n)S_{\mathrm{approx}}\left(\frac{\partial J}{\partial \mathbf{W}}\mathbf{W}^\mathrm{T}(n)\mathbf{W}(n)\right), \tag{6.46}$$

keeping in mind that only the dth row ($S_{\mathrm{approx}} = S_d$) or Lth column ($S_{\mathrm{approx}} = S^L$) of $\Delta\mathbf{W}(n)$ has to be computed. It may be difficult to interpret the meaning of these approximations. What we can say is that certain gradient terms $\partial J/\partial w_{ij}^{nm}$ which do not belong to the Sylvester subspace are involved in the computation of the natural gradient. We note that the two typical choices S_1 and S^L shown in Fig. 6.1 are obtained as special cases. They were discussed by Aichner et al. with regard to the self-closedness and to the causality of the separation system [5].

Implementation of the NG-SOS-BSS updates

In the following we provide a precise implementation of the NG-SOS-BSS algorithm (5.41). We first need estimates for the output cross correlations $\mathbf{E}\{y_n(p)y_m(p-\tau)\}$ for $n, m = 1, \ldots, N$ and $\tau = -L+1, \ldots, L-1$. A usual (biased) estimation $r_{y_n y_m, \tau}(p)$ of $\mathbf{E}\left\{y_n(p)y_m(p-\tau)\right\}$ is obtained by averaging over blocks of length L and can be written as

$$r_{y_n y_m, \tau}(p) = \sum_{\kappa=0}^{L-1-\tau} y_n(p-\kappa)y_m(p-\kappa-\tau). \qquad (6.47)$$

The estimator in (6.47) sums up L sample products $y_n(p-\kappa)y_m(p-\kappa-\tau)$ for the delay $\tau = 0$ whereas the estimate consists in one sample product for $\tau = L-1$, which results in a larger estimation variance. Since the estimates $r_{y_n y_m, \tau}$ are all normalized by the same power term $\|\mathbf{y}_n\|^2$, the estimates $r_{y_n y_m, \tau}$ for a larger delay (τ close to L) are eventually weighted down with respect to $r_{y_n y_m, \tau}$ for τ close to 0. We note that other (unbiased) correlation estimators may be considered, for example, by summing on the same number of sample products $y_n(p-\kappa)y_m(p-\kappa-\tau)$ for each delay τ. The estimator (6.47) has the advantage to require the current output blocks on L points only.

The output signal power is regularized using a parameter $\alpha \in [0,1]$ as shown in (7.26) (more details in Sect. 7.1.2). We set

$$\tilde{r}_{y_n y_n, 0}(p) = \begin{cases} r_{y_n y_n, 0}(p) & \text{if } r_{y_n y_n, 0}(p) > 2\alpha, \\ \frac{1}{2}\left(\alpha + r_{y_n y_n, 0}(p)\right) & \text{otherwise.} \end{cases} \qquad (6.48)$$

We also need the $L \times Q$ projection matrix $\mathbf{P}^{L \times Q}_{[l_0, l_0+L-1]}$ which is defined as follows:

$$\left[\mathbf{P}^{L \times Q}_{[l_0, l_0+L-1]}\right]_{ij} = \begin{cases} 1 & \text{if } j \in [l_0, l_0 + L - 1] \quad \text{and} \quad i = j + l_0, \\ 0 & \text{otherwise.} \end{cases} \qquad (6.49)$$

Self-closed update

Let us introduce the $L \times 1$ output cross-correlation vectors $\mathbf{r}^{(d)}_{\mathbf{y}_n \mathbf{y}_p}(p)$ for $n, p = 1, \ldots, N$:

$$\mathbf{r}^{(d)}_{\mathbf{y}_n \mathbf{y}_p}(p) = \left(r_{y_n y_p, -d}(p), \ldots, r_{y_n y_p, L-1-d}(p)\right)^{\mathrm{T}}. \qquad (6.50)$$

Substituting S_{d+1} for S_{approx} in (6.46) and reworking the NG-SOS-BSS equation (5.41), we can derive

$$\boxed{\Delta \mathbf{w}_{nm} = -\mu \sum_{k=1}^{K} \sum_{\substack{p=1 \\ p \neq n}}^{N} \mathbf{P}^{L \times 2L-1}_{[d, L+d-1]} \left(\mathbf{w}_{pm} * \mathbf{r}^{(d)}_{\mathbf{y}_n \mathbf{y}_p}(kL)\right) / \tilde{r}_{y_n y_n, 0}(kL).} \qquad (6.51)$$

Different values of d yield different update rules.

Non-self-closed update

Let us introduce the $2L - 1 \times 1$ output cross-correlation vectors $\mathbf{r}_{\mathbf{y}_n \mathbf{y}_p}(p)$ for $n, p = 1, \ldots, N$:

$$\mathbf{r}_{\mathbf{y}_n \mathbf{y}_p}(p) = \left(r_{y_n y_p, -L+1}(p), \ldots, r_{y_n y_p, L-1}(p) \right)^{\mathrm{T}}. \tag{6.52}$$

Substituting S^L for S_{approx} in (6.46) and reworking the NG-SOS-BSS equation (5.41), we can derive

$$\Delta\mathbf{w}_{nm} = -\mu \sum_{k=1}^{K} \sum_{\substack{p=1 \\ p \neq n}}^{N} \mathbf{P}_{[L-1,2L-2]}^{L \times 3L-2} \left(\mathbf{w}_{pm} * \mathbf{r}_{\mathbf{y}_n \mathbf{y}_p}(kL) \right) / \tilde{r}_{y_n y_n, 0}(kL). \tag{6.53}$$

Partial BSS

The PBSS update (5.52) can be implemented similarly. In the case of $M = 4$, $N = 2$ for the self-closed update, this yields:

$$\Delta\mathbf{w}_{nm} = -\mu \sum_{k=1}^{K} \sum_{p \in \mathcal{K}_n} \mathbf{P}_{[d,L+d-1]}^{L \times 2L-1} \left(\mathbf{w}_{pm} * \mathbf{r}_{\mathbf{y}_n \mathbf{y}_p}^{(d)}(kL) \right) / \tilde{r}_{y_n y_n, 0}(kL)$$

with $\mathcal{K}_1 = \{2, 3, 4\}$ and $\mathcal{K}_n = \{1\}$ for $n = 2, 3, 4$.
$$\tag{6.54}$$

The non-self-closed updates are given by:

$$\Delta\mathbf{w}_{nm} = -\mu \sum_{k=1}^{K} \sum_{p \in \mathcal{K}_n} \mathbf{P}_{[L-1,2L-2]}^{L \times 3L-2} \left(\mathbf{w}_{pm} * \mathbf{r}_{\mathbf{y}_n \mathbf{y}_p}(kL) \right) / \tilde{r}_{y_n y_n, 0}(kL). \tag{6.55}$$

As mentioned in Sect. 5.4, we consider not adapting the filters $\mathbf{w}_{nm}$ for $n, m > 1$ and $n \neq m$. We could experimentally observe that this does not impair the separation performance, while reducing the amount of computation significantly.

The pseudocode for the self-closed BSS and PBSS algorithms is summarized in Table 6.1. The non-self-closed algorithms are implemented similarly, replacing $\mathbf{r}_{\mathbf{y}_n \mathbf{y}_p}^{(d)}(p)$ with $\mathbf{r}_{\mathbf{y}_n \mathbf{y}_p}(p)$ in (6.52) at line 5 and using (6.53) and (6.55) at lines 9 and 10, respectively.

6.1.6 Discussion: Which Natural Gradient is Best?

We have shown in Sect. 6.1.5 how the natural gradient can be implemented in the time domain with filters of finite length L. Now, we are faced with the choice of a particular type of natural gradient (self-closed or non-self-closed) and of the parameter $d \in \{0, \ldots, L-1\}$. In this section, we attempt to discuss the implications of this choice.

Self-closedness

Constraining the term $\overline{\mathbf{D}}$ in (6.27) seems more restrictive in the set $\overline{S}$ than in the larger set $\overline{T}$. Consequently, one could expect better performance with the

Table 6.1. Pseudocode for the self-closed batch algorithms BSS and PBSS in (6.51) and in (6.54). The non-self-closed algorithms are implemented similarly, replacing $\mathbf{r}_{\mathbf{y}_n\mathbf{y}_p}^{(d)}(p)$ with $\mathbf{r}_{\mathbf{y}_n\mathbf{y}_p}(p)$ in (6.52) at line 5 and using (6.53) and (6.55) at lines 9 and 10, respectively

---

Input

$\mathbf{x}(kL)$ for $k = 1, \ldots, K$
Note: a higher frame rate can be used with $\mathbf{x}(k\beta L)$ for $k = 1, \ldots, K$, choosing β so that $\beta L \in \{1, \ldots, L\}$.

Parameters

$d \in \{0, \ldots, L-1\}$	acausal filter length (delay)
μ	step-size
α	regularization parameter
N_{iter}	number of iterations
$\mathbf{w}_{nm}(0)$ for $n, m = 1, \ldots, M$	initial separation filters

define $\mathcal{I} = \{(m, m) | m = 1, \ldots, M\}$

define $\mathcal{J} = \begin{cases} \{(1,2), (2,1)\} & \text{for BSS and } M = 2 \\ \{(1,2), (1,3), (1,4), (2,1), (3,1), (4,1)\} & \text{for PBSS and } M = 4 \end{cases}$

Computations

---

1. for $n_{\text{iter}} = 1, \ldots, N_{\text{iter}}$
2. for $k = 1, \ldots, K$
3. compute the output vector (using convolutions):
$$\mathbf{y}(kL) = \mathbf{W}(n_{\text{iter}} - 1)\mathbf{x}(kL)$$
4. for $(n, m) \in \mathcal{J}$:
5. compute the vector $\mathbf{r}_{\mathbf{y}_n\mathbf{y}_m}^{(d)}(kL)$ according to (6.47) and (6.50)
6. compute the power $r_{y_n y_n, 0}(kL)$ according to (6.47)
7. compute the regularized power $\tilde{r}_{y_n y_n, 0}(p)$ according to (6.48)
8. for $(n, m) \in \mathcal{J} \cup \mathcal{I}$:
9. in the case of BSS, compute $\Delta \mathbf{w}_{nm}$ according to (6.51):
$$\Delta \mathbf{w}_{nm} = -\mu \sum_{k=1}^{K} \sum_{\substack{p=1 \\ p \neq n}}^{N} \mathbf{P}_{[d, L+d-1]}^{L \times 2L-1} \left(\mathbf{w}_{pm}(n_{\text{iter}} - 1) * \mathbf{r}_{\mathbf{y}_n\mathbf{y}_p}^{(d)}(kL) \right) / \tilde{r}_{y_n y_n, 0}(kL)$$
10. in the case of PBSS, compute $\Delta \mathbf{w}_{nm}$ according to (6.54):
$$\Delta \mathbf{w}_{nm} = -\mu \sum_{k=1}^{K} \sum_{p \in \mathcal{K}_n} \mathbf{P}_{[d, L+d-1]}^{L \times 2L-1} \left(\mathbf{w}_{pm}(n_{\text{iter}} - 1) * \mathbf{r}_{\mathbf{y}_n\mathbf{y}_p}^{(d)}(kL) \right) / \tilde{r}_{y_n y_n, 0}(kL)$$
11. $\mathbf{w}_{mn}(n_{\text{iter}}) = \mathbf{w}_{mn}(n_{\text{iter}} - 1) + \Delta \mathbf{w}_{nm}$

Output

$\mathbf{W}(N_{\text{iter}})$	separation system
$\mathbf{y}(kL)$ for $k = 1, \ldots, K$	output signals

---

non-self-closed update in (6.38) than with the self-closed update in (6.37). In the literature on second-order statistics BSS algorithms for acoustic mixtures, it is rarely clear whether the proposed updates are self-closed or not, hence the difficulty of concluding anything from published works. Furthermore, because of the lack of knowledge on the convergence behavior of BSS algorithms, it is not possible to make a theoretical statement about whether or not the self-closedness improves performance. For this reason, we will compare self-closed and non-self-closed updates experimentally in Sect. 6.3.

Special case $d = 0$

Since d represents the number of coefficients which are treated as acausal in the convolution $\star_d$, we call d the acausal length of the separation system. For $d = 0$, the separation system adapted by (6.56) is called a "causal" system while it is called "acausal" for $d > 0$. (Of course, the individual filters $\mathbf{w}_{nm}$ are still causal, that is, $w_{nm,l} = 0$ for all $l < 0$.)

Let us denote the separation system $\overline{\mathbf{W}}(z)$ at iteration step n by $\overline{\mathbf{W}}(z, n)$. If we set $d = 0$ in (6.27), the adaptation can be written in the z-domain as

$$\overline{\mathbf{W}}(z, n+1) = \overline{\mathbf{W}}(z, n) + \overline{\mathbf{D}}(z, n) \star_0 \overline{\mathbf{W}}(z, n), \qquad (6.56)$$

where $\overline{\mathbf{D}}(z, n) \in \overline{S}$. The set $\overline{S}$ defines a group with the operation $\star_0$ [95]. This implies several properties for the self-closed adaptation rule. (It should be noted that the mixing/separation systems also have a group structure in the two limit cases $L = 1$ and $L = +\infty$ [6, 25].)

The *equivariance* property guarantees a uniform convergence behavior for any mixing matrix. Just as we considered the z-transform of the separation matrix $\overline{\mathbf{W}}(z, n)$, we can consider the z-transform of the mixing matrix $\overline{\mathbf{H}}(z)$. The global system, truncated to its first L taps, is given by $\overline{\mathbf{C}}(z, n) = \overline{\mathbf{W}}(z, n) \star_0 \overline{\mathbf{H}}(z)$. Since $\star_0$ is associative, we have

$$\mathbf{\Delta}\overline{\mathbf{W}}(z, n) \star_0 \overline{\mathbf{H}}(z), = \left(\overline{\mathbf{D}}(z, n) \star_0 \overline{\mathbf{W}}(z, n)\right) \star_0 \overline{\mathbf{H}}(z), \qquad (6.57)$$

$$= \overline{\mathbf{D}}(z, n) \star_0 \left(\overline{\mathbf{W}}(z, n) \star_0 \overline{\mathbf{H}}(z)\right), \qquad (6.58)$$

$$= \overline{\mathbf{D}}(z, n) \star_0 \overline{\mathbf{C}}(z, n). \qquad (6.59)$$

Therefore, postmultiplying (6.56) by $\overline{\mathbf{H}}$ yields

$$\overline{\mathbf{C}}(z, n+1) = \overline{\mathbf{C}}(z, n) + \overline{\mathbf{D}}(z, n) \star_0 \overline{\mathbf{C}}(z, n). \qquad (6.60)$$

Note that $\overline{\mathbf{D}}(z, n)$ depends only on the output signals. Therefore, the trajectory of the global system $\overline{\mathbf{C}}(z, n)$ in (6.60) depends only on the initial point $\overline{\mathbf{C}}(z, 0)$ and on the realization of the *source* signals. This indicates that the convergence of the algorithm depends on a particular mixing matrix $\overline{\mathbf{H}}(z)$ only through the initial $\overline{\mathbf{C}}(z, 0)$. Note that for $d > 0$, the operation $\star_d$ is not associative: The equivariance property holds only for the self-closed update and $d = 0$.

Another property of $\star_0$ is that the adaptation of any given separation filter does not interfere with the other filters.[7] To illustrate this property, let us consider the two-source two-sensor case ($M = N = 2$). As can be seen from (6.50)

---

[7] It may be noticed that the SAD algorithm relies on this property.

and (6.51) for $d = 0$, the computation of $\mathbf{\Delta w}_{1m}$ for $m = 1, 2$ involves the output cross correlations $\mathbf{E}\{y_1(p)y_2(p - \tau)\}$ only for positive delays $\tau \geq 0$. Similarly, the computation of $\mathbf{\Delta w}_{2m}$ includes only $\mathbf{E}\{y_1(p)y_2(p - \tau)\}$ for negative delays $\tau \leq 0$. Therefore, $\mathbf{w}_{1m}$ and $\mathbf{w}_{2m}$ for $m = 1, 2$ are driven by different cross-correlation terms, except the zero-lag correlation $\mathbf{E}\{y_1(p)y_2(p)\}$ that is involved in both $\mathbf{\Delta w}_{1m}$ and $\mathbf{\Delta w}_{2m}$, for $m = 1, 2$. By contrast, if $d = L/2$ or with the non-self-closed update (6.53), $\mathbf{\Delta w}_{1m}$ and $\mathbf{\Delta w}_{2m}, m = 1, 2$ have L output cross-correlation terms in common.

Unfortunately, the utilization of self-closed update $\star_0$ also has some restrictions. It can be shown that if $\overline{\mathbf{D}}(z, n) \in \overline{\mathbf{S}}$, then according to [95] we have

$$\left[\overline{\mathbf{W}}(z, n)\right]_0 = \mathbf{0} \Leftrightarrow \left[\overline{\mathbf{D}}(z, n) \star_0 \overline{\mathbf{W}}(z, n)\right]_0 = \mathbf{0}. \tag{6.61}$$

Therefore, it is necessary to initialize the separation system $\overline{\mathbf{W}}(z, n)$ with $\left[\overline{\mathbf{W}}(z, 0)\right]_0 \neq \mathbf{0}$ in (6.56), since otherwise the first tap of the separation filters would remain zero.[8] In practice, we initialize with unit responses, $\overline{\mathbf{W}}(z, 0) = \mathbf{I}$. The update (6.56) is inappropriate if we want to initialize the separation system with a delay, for example with $\overline{\mathbf{W}}(z, 0) = \mathbf{I}z^{-L/2}$. Such an initialization may be useful for example in the scenario depicted in Fig. 6.2b, where the sources are placed on the same side of the microphone median plane. By contrast, update (6.56) is appropriate when the sources are placed on both sides of the median plane (Fig. 6.2a) [5]. In this scenario, the input signals do not need to be delayed. In other words, a causal separation system can perform the separation.

Note that in the two-source two-sensor scenario depicted in Fig. 6.2a, the source signals $s_1(p)$ and $s_2(p)$ *cannot* be canceled at the outputs $y_1(p)$ and $y_2(p)$, respectively, if the filters $\mathbf{w}_{11}$ and $\mathbf{w}_{22}$ do not delay the input signals $x_1(p)$ and $x_2(p)$. Therefore, the update (6.56) neutralizes the permutation ambiguity. Conversely for $d = L - 1$, the source signals $s_1(p)$ and $s_2(p)$ cannot be canceled at the outputs $y_2(p)$ and $y_1(p)$, respectively. In this case, the updates (6.53) and (6.51) should converge to a permuted separating system.

How does the causality of the separation system affect the convergence of the separation algorithm? A theoretical answer to this question does not yet exist. Nevertheless, the influence of causality in source separation has been reported in various contexts [34, 79]. For example, Douglas et al. showed in a simplified analysis that separation algorithms exhibit worse performance for acausal separation systems (i.e., for $d > 0$) [34].

---

[8] According to (6.61), if the first tap of the initial separation system is nonzero, then we have $\left[\overline{\mathbf{W}}(z, n)\right]_0 \neq \mathbf{0}$ for all $n > 0$. In a number of simulations, we could observe that the initialization $\overline{\mathbf{W}}(z, 0) = \mathbf{I}$ yields diagonal filters that are not only nonzero but also minimum phased. Unfortunately, we have no explanation to offer for this observation, that may be limited to the simulation setup and may have no generality.

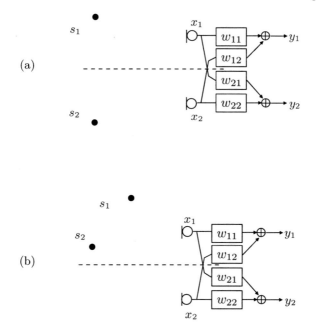

Fig. 6.2. (a) The sources are apart from the microphones median plan and a causal separation system is sufficient. (b) The sources are in the same half-plane and an acausal separation system is necessary

6.2 Online Adaptation

The framework presented so far provides offline NG-SOS-BSS algorithms: It is assumed that the observed input $\mathbf{x}(p)$ is available as a batch of T samples $p = 1, \ldots, T$. In reality, we have to cope with a continuously growing number of samples $\mathbf{x}(1), \mathbf{x}(2), \ldots$, since new observations keep on coming during the course of the iterations. Moreover, the source signals spectra are continuously changing. The separation algorithm should be able to track these changes online.

6.2.1 Block-Wise Batch Adaptation

It is not difficult to transform offline algorithms into online algorithms with block-wise batch processing. After a block of βL new samples have been received, we store them in a batch buffer $\mathcal{B}$ that contains a certain number of last input samples. This batch is processed offline for, say, N_{iter} iterations, until new samples are processed. This general approach is summarized in Table 6.2. Several variations on this scheme exist, see, e.g., [4, 73].

Different parameters need to be defined by the user for the implementation of a block-wise batch algorithm:

Table 6.2. Block-wise batch processing scheme with parameters $K, N_{iter} \geq 1$ and β so that $1 \leq \beta L \leq (K+1)L - 1$. In $\mathbf{W}(n,p)$, the first argument n denotes the iteration index and the second argument p denotes the time index

0.	Initialize $\mathbf{W}(0,0)$, set the frame counter to $l = 1$.
1.	Acquire βL new samples $x_m(p), m = 1, \ldots, M$. Store them in the batch buffer $\mathcal{B}$ and discard the oldest βL input samples from $\mathcal{B}$ so that $\mathcal{B} = \{x_m(\tau), \tau = p - (K-1)L + 2, \ldots, p,$ for $m = 1, \ldots, M\}$.
2.	Set the last computed $\mathbf{W}(N_{\text{iter}}, (l-1)\beta L)$ as initial separation system: $\mathbf{W}(0, l\beta L) = \mathbf{W}(N_{\text{iter}}, (l-1)\beta L)$. Run a batch BSS algorithm on $\mathcal{B}$ as described in Table 6.1 with N_{iter} iterations. This provides $\mathbf{W}(N_{\text{iter}}, l\beta L)$.
3. (optional)	The output samples might have been delivered at step 2. Otherwise, compute the output samples that correspond to the last βL input samples. If $\beta = 1$, this is simply $\mathbf{y}(lL) = \mathbf{W}(N_{\text{iter}}, lL)\mathbf{x}(lL)$.
4.	Increment the frame counter, $l \leftarrow l + 1$. Repeat from step 1.

- The number of blocks K in the batch $\mathcal{B}$ is required. K is the number of output correlation matrices $\mathbf{R_{yy}}(kL), k = 1, \ldots, K$ that should be jointly diagonalized by the batch algorithm. K should increase the speed of convergence. It also increases the computational cost and the memory requirement.
- The maximal number of iterations N_{iter} needs to be set. N_{iter} influences the speed of convergence and the computational cost.
- The parameter β is set so that βL is an integer value. βL controls the number of new input samples that are acquired before running the batch algorithm on $\mathcal{B}$. Since the number of new input samples must not be larger than the buffer size, β is chosen so that $1 \leq \beta L \leq (K+1)L - 1$. A small value of β entails better tracking capability, higher computational cost and lower memory requirement. The number of batch runs per second is $f_s/\beta L$.

The influence that these parameters may have on the performance is examined experimentally in Sect. 8.4.

6.2.2 Sample-Wise Adaptation

By setting $\beta L = 1$, the block-wise batch approach in Sect. 6.2.1 may be performed for each new arriving input sample. This increases the amount of computation[9] significantly. Fortunately for $N = 2$ and if we do not adapt the

---

[9] Using the results of Sect. 8.2.2, one can estimate that the number of real operations (multiplications and additions) for a sample-by-sample-updated four-input

diagonal filters, the block-wise approach may be approximated so that sample-by-sample updating and rapid tracking become feasible. The following derives this sample-by-sample version.

Let us start with the block-wise batch approach applied to the NG-SOS-BSS generic formula (5.41) with the parameters $\beta L = K = N_{\text{iter}} = 1$:

$$\mathbf{W}(p) = \mathbf{W}(p-1) - \mu \, \text{diag}^{-1}\left(\mathbf{R_{yy}}(p)\right) \text{boff}\left(\mathbf{R_{yy}}(p)\right) \mathbf{W}(p). \quad (6.62)$$

To estimate the output correlation matrix $\mathbf{R_{yy}}(p)$, the output vector $\mathbf{y}(p) = \mathbf{W}(p)\mathbf{x}(p)$ is evaluated, i.e., the outputs on the past L samples $y_n(p-L+1), \ldots, y_n(p)$ for $n = 1, \ldots, N$ are evaluated. Note that among these L output samples, $L-1$ have already been evaluated from the previous iteration using $\mathbf{W}(p-1)$, namely $y_n(p-L+1), \ldots, y_n(p-1)$. Assuming that the separation system $\mathbf{W}(p)$ changes slowly, we only need to compute the last sample $y_n(p)$, for $n = 1, \ldots, N$. Since $\mathbf{y}(p)$ contains $L-1$ past samples, the underlying approximation is in fact

$$\mathbf{W}(p) \approx \mathbf{W}(p-1) \approx \mathbf{W}(p-2) \approx \ldots \approx \mathbf{W}(p-L+1). \quad (6.63)$$

This reduces the computational complexity to $\mathcal{O}(L)$, whereas the computation of $\mathbf{y}(p)$ with implementation of the convolution in the DFT-domain has complexity $\mathcal{O}(L \log_2 L)$ (see Sect. 8.2).

Next, we consider the estimation of $\mathbf{R_{yy}}(p)$ itself. The simplest estimation of $\text{boff}\left(\mathbf{R_{yy}}(p)\right)$ is obtained with the instantaneous estimate

$$\text{boff}\left(\widehat{\mathbf{R}}_{\mathbf{yy}}(p)\right) = \text{boff}\left(\mathbf{y}(p)\mathbf{y}^{\mathrm{T}}(p)\right). \quad (6.64)$$

The straightforward implementation of the product $\text{boff}\left(\mathbf{y}(p)\mathbf{y}^{\mathrm{T}}(p)\right)$ has a complexity of $\mathcal{O}(L^2)$. Fortunately, since only the last sample $y_n(p)$ is new in the output vector $\mathbf{y}_n(p)$, we can exploit the fact that

$$\widehat{\mathbf{R}}_{\mathbf{y}_i\mathbf{y}_j}(p) = \begin{bmatrix} y_i(p)y_j(p) & \cdots & y_i(p)y_j(p-L+1) \\ \vdots & \widehat{\mathbf{R}}_{\mathbf{y}_i\mathbf{y}_j}^{[L-1\times L-1]}(p-1) & \\ y_i(p-L+1)y_j(p) & & \end{bmatrix}. \quad (6.65)$$

$\widehat{\mathbf{R}}_{\mathbf{y}_i\mathbf{y}_j}^{[L-1\times L-1]}(p-1)$ is constructed from $\widehat{\mathbf{R}}_{\mathbf{y}_i\mathbf{y}_j}(p-1)$ as follows: For an $L \times L$ matrix $\mathbf{A}$, $\mathbf{B} = \mathbf{A}^{[L_1 \times L_2]}$ is the $L_1 \times L_2$ matrix that contains the same elements as $\mathbf{A}$ in its first L_1 rows and first L_2 columns, i.e.,

$$[\mathbf{B}]_{ij} = [\mathbf{A}]_{ij} \quad \forall i = 1, \ldots, L_1, \, j = 1, \ldots, L_2. \quad (6.66)$$

---

two-output separation system with $f_{\mathrm{s}} = 16\,\text{kHz}$ sampling frequency, $L = 256$ and $K = N_{\text{iter}} = 1$ is about 15,000 MFLOPS (million of floating point operation per second). This amount of computation may be realized by high-end digital signal processors (DSP), however, for current consumer-device DSPs, the performance typically ranges from 100 to 1,000 MFLOPS (see for example the specifications of the TMS320C54x DSP family, available on www.ti.com).

Consequently, the instantaneous estimation of boff $(\mathbf{R_{yy}}(p))$ can be performed with complexity $\mathcal{O}(L)$.

Finally, we have to implement the convolutions represented by the matrix product boff $(\mathbf{R_{yy}}(p))\,\mathbf{W}(p)$. A straightforward realization using the fast-Fourier transform (FFT) has a complexity of $\mathcal{O}(L\log_2 L)$. Fortunately, there is a way to avoid this operation in the case of two sources $(N = 2)$. Let us additionally constrain the diagonal filters to unit responses, that is, $\mathbf{w}_{nn} = \boldsymbol{\delta}_d, n = 1, 2$. Only the filters $\mathbf{w}_{12}$ and $\mathbf{w}_{21}$ are adapted. We can easily verify that the product boff $(\mathbf{R_{yy}}(p))\,\mathbf{W}(p)$ involves convolutions of the output cross correlations with unit filters only. These convolutions do not need to be carried out. Estimating the output power by $r_{nn,0}(p) = \|\mathbf{y}_n(p)\|^2$ and using the regularization with parameter $\alpha \in [0, 1]$ yields

$$\mathbf{w}_{nm}(p+1) = \mathbf{w}_{nm}(p) - \mu \frac{y_n(p-d)\mathbf{y}_m(p)}{\tilde{r}_{y_n y_n, 0}(p)}, \tag{6.67}$$

for $(n, m) \in \{(2, 1), (2, 1)\}$ and with

$$\tilde{r}_{y_n y_n, 0}(p) = \begin{cases} \|\mathbf{y}_n(p)\|^2 & \text{if } \|\mathbf{y}_n(p)\|^2 > 2\alpha, \\ \frac{1}{2}\left(\|\mathbf{y}_n(p)\|^2 + \alpha\right) & \text{otherwise.} \end{cases}$$

Interestingly, the constraint $\mathbf{w}_{nn} = \boldsymbol{\delta}_d, n = 1, 2$ not only decreases the number of filters $\mathbf{w}_{nn'}$ to be adapted from 4 to 2 but also reduces the complexity from $\mathcal{O}(L\log_2 L)$ to $\mathcal{O}(L)$. That is, the complexity of the sample-wise adapted NG-SOS-BSS algorithm in (6.67) is the same as that of the NLMS algorithm (4.7). Note that this does *not* hold for $N > 2$. However, the sample-wise adapted NG-SOS-BSS algorithm in (6.67) may be extended to partial BSS (PBSS). In the case $N = 2$, the sample-wise online algorithm is

$$\mathbf{w}_{nm}(p+1) = \mathbf{w}_{nm}(p) - \mu \frac{y_n(p-d)\mathbf{y}_m(p)}{\tilde{r}_{y_n y_n, 0}(p)}, \tag{6.68}$$

$$\text{for } (n, m) \in \{(2, 1), \dots, (M, 1), (1, 2), \dots, (1, M)\}$$

$$\text{with } \tilde{r}_{y_n y_n, 0}(p) = \begin{cases} \|\mathbf{y}_n(p)\|^2 & \text{if } \|\mathbf{y}_n(p)\|^2 > 2\alpha, \\ \frac{1}{2}\left(\|\mathbf{y}_n(p)\|^2 + \alpha\right) & \text{otherwise.} \end{cases}$$

6.3 Experimental Results

This section investigates the performance of NG-SOS-BSS algorithms experimentally. It was not possible to determine theoretically whether or not the self-closed updates perform better than their non-self-closed counterparts. Based on experimental results, this section decides which updates will be used in the following of this book.

Experimental conditions

The experiments are conducted with real recordings performed in a car cabin. The experimental setup is described in Appendix A. The speech source signals are emitted from the driver and codriver positions. For results in batch mode, the recordings are carried out using *artificial heads*. Hence, the mixing system may be reasonably assumed time-invariant. The 10-s driver and codriver signals consist of male and female voices, respectively.

Parameter settings

The adaptation parameters α and μ were optimized for maximum convergence speed and stability. The artificial head recordings have been used to tune the parameters μ and α for the online algorithms, which we finally set to $\mu = \alpha = 0.002$ for all online NG-SOS-BSS algorithms. It should be mentioned that setting μ larger than 0.002 may lead to larger SIR improvement. However, we observed that this leads to stability problems for certain source signals. We have noticed that the choice of the parameters μ and α may be sensitive. When one or several sources are silent, the output power may become very small. In this case, in spite of the regularization scheme (7.26), the adaptation may become unstable. (Recollect that (7.26) was derived from an analysis of a simpler instantaneous decorrelation algorithm.) More robustness may be obtained by choosing a larger α, at the price of a lower performance. Consequently, we selected $\mu = \alpha = 0.002$ as a trade-off between SIR improvement and stability. The online algorithms have been tested on the same speech signals as in Sect. 4.5, which are recordings of *real speakers*. Optionally, background noise has been added to these recordings, as in Sect. 4.5. In all experiments, the filter length is set to $L = 256$ and the sampling frequency is $f_s = 16\,\mathrm{kHz}$. The filters were initialized with $\overline{\mathbf{W}}(z, 0) = \mathbf{I}z^{-d}$, that is,

$$\mathbf{w}_{nm} = \boldsymbol{\delta}_d. \tag{6.69}$$

The parameter settings are summarized in Table 6.4 for the four-element compact array mounted in the rear-view mirror and in Table 6.3 for the two-element distributed array mounted on the car ceiling.

Performance measure

The separation performance is quantified by the SIR improvement as defined in Sects. 2.4.1 and 2.4.2.

6.3.1 Experiments with the Four-Element Compact Array Mounted in the Rear-View Mirror

The four-element compact array mounted in the rear-view mirror is depicted in Fig. A.2.

Table 6.3. Online NG-SOS-BSS algorithm short reference table with the parameter settings for the two-element distributed array mounted on the car ceiling shown in Fig. A.3. The filter length is set to $L = 256$ at $f_s = 16$ kHz

	block-wise	sample-wise
structure	MIMO (see Fig. 5.1b)	MIMO (see Fig. 5.1c) unit diagonal filters
section	6.2.1	6.2.2
equation	(6.51) and (6.53)	(6.67)
step-size	$\mu = 0.002$ $\alpha = 0.002$	
other parameters	$K = \beta = N_{\text{iter}} = 1$	

Table 6.4. Online NG-SOS-BSS algorithm short reference table with the parameter settings for the four-element compact array mounted in the rear-view mirror shown in Fig. A.2. The filter length is set to $L = 256$ at $f_s = 16$ kHz

	block-wise	sample-wise
structure	MIMO as in (5.53)	(see Fig. 5.3c) unit diagonal filters
section	6.2.1	6.2.2
equation	(6.54) and (6.55)	(6.68)
step-size	$\mu = 0.002$ $\alpha = 0.002$	
other parameters	$K = \beta = N_{\text{iter}} = 1$	

Batch Algorithms

The question of the self-closedness could not be answered theoretically. For this reason, we will experimentally compare the self-closed with the non-self-closed updates using batch algorithms.

Choice of d

A first parameter of interest is the acausal length d of the separation filters, on which the self-closed update depends. To evaluate how the performance depends on d, we evaluate the SIR improvement after 100 iterations of the update in (6.51). Figure 6.3 shows the results. In the case of the four-element compact array mounted in the rear-view mirror, it can be seen that the performance slowly increases until it attains its maximum around $d = 50$. This indicates that, even though the source microphone does not require an acausal separation system, the best separation performance is obtained if we allow the separation system to have an acausal part. This may be explained by the fact that, as in the case of LCMV beamformers, the separation system identifies ratios of acoustic transfer functions, which have a significant acausal part [50].

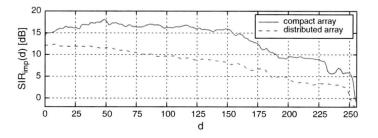

Fig. 6.3. Performance of the self-closed block-wise update as a function of d. For each value of $d = 0, \ldots, L-1$, the SIR improvement after $n = 100$ iterations is shown

For $d \geq L/2$, that is when the acausal part is longer than the causal part of the separation system, the performance significantly decreases. As indicated in Sect. 6.1.6, updates (6.51) and (6.53) should converge to a permuted separating system for a separation system with a long acausal part. This is rather ineffective with the initial separation system (6.69), since the input SIRs are greater than zero (about $2\,$dB). For our next experiments, we retain $d = 0$ and $d = 50$.

Comparison of self-closed and non-self-closed updates

We now apply PBSS algorithms (6.54) and (6.55) as described in Table 6.1. For the sake of comparison, the results obtained using the two outermost microphones are also given. With speech signals, the wavelength of interest is large relative to the aperture of the four-element compact array mounted in the rear-view mirror. Hence, selecting the two outermost microphones may be considered as a rough preprocessing to reduce the dimension of the observed signals from $M = 4$ to $M = 2$ while preserving the aperture of the microphone array. In this case, the adaptation is performed using the NG-SOS-BSS algorithms (6.51) and (6.53). The results are shown in Fig. 6.4. The following comments can be made.

- The self-closed update leads to the best results. The non-self-closed update rule may lead to an early saturation and seems less stable. As observed in [5], we find that the self-closed update rules are more robust. This might be expected from the "good" properties of the self-closed update in the case $d = 0$ given in Sect. 6.1.6.
- The performance near convergence is slightly better with an acausal length of $d = 50$ taps, even if the source-microphone arrangement does not require an acausal length, as depicted in Fig. 6.2a. This coincides with the results shown in Fig. 6.3.
- We obtain the best performance using $M = 4$ microphones. In terms of performance near convergence, the additional microphones bring about $4\,$dB improvement relative to the performance for the two outermost microphones.

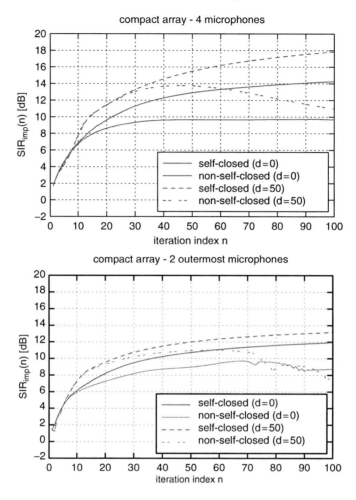

Fig. 6.4. Comparison of batch NG-SOS-BSS algorithms. *Upper plot*: SIR improvement for $M = 4$ microphones using PBSS algorithms (6.54) and (6.55). *Lower plot*: SIR improvement for the $M = 2$ outermost microphones and BSS algorithms (6.51) and (6.53)

Online algorithms

This section examines the performance of online NG-SOS-BSS algorithms. Only the self-closed update algorithm (6.54) with $M = 4$ microphones is considered, because it leads to the best results in batch mode. The adaptation is performed block-wise as described in Table 6.2 with nonoverlapping blocks ($K = \beta = N_{\text{iter}} = 1$). The sample-wise algorithm (6.68) is also considered.

The results are given in terms of interference and target signal level reductions in Fig. 6.5. The following comments may be made:

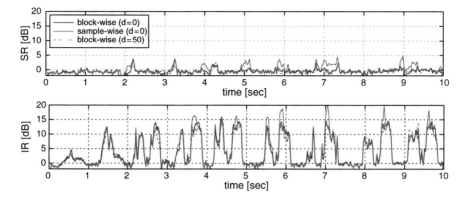

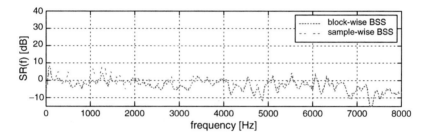

Fig. 6.5. Comparison of online PBSS algorithms for the four-element compact array mounted in the rear-view mirror. Block-wise algorithms are based on the self-closed updates (6.54) with $K = \beta = N_{\text{iter}} = 1$. The sample-wise update is given by (6.68)

Fig. 6.6. Reduction of the desired signal level as a function of the frequency with the four-element compact array mounted in the rear-view mirror (no background noise)

- The performance of the two block-wise batch NG-SOS-BSS algorithms considered here are very similar. They provide a peak reduction of the codriver signal level of about 15 dB.
- The best peak performance is attained by the sample-wise algorithm (6.68) because of its tracking capability. However, it seems to reduce the target signal level more than its block-wise counterpart.

The reduction of the desired signal level as a function of the frequency is illustrated in Fig. 6.6. The sample-wise adapted algorithm seems to cause a slightly higher distortion of the desired signal at low frequencies ($<2,000\,\text{Hz}$). Compared to the distortion caused by the DTD-NLMS algorithm (see Fig. 4.8), the distortion caused by NG-SOS-BSS algorithms seems slightly higher. However, the distortion caused by NG-SOS-BSS algorithms is significantly smaller than the distortion caused by ILMS algorithms, in particular at frequencies $f < 1,000\,\text{Hz}$ where speech signals have most

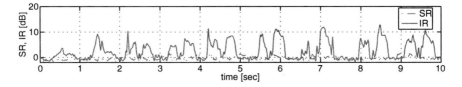

Fig. 6.7. Online PBSS with road noise recorded at $100 \, \text{km/h}^{-1}$, the signal-to-noise ratio at the microphones amounts to 10–15 dB. The adaptation is performed using the self-closed block-wise algorithm (6.54) and $d = 0$, $K = \beta = N_{\text{iter}} = 1$

energy. These results have also been confirmed through subjective listening tests.

It can be seen in Fig. 6.7 that the self-closed NG-SOS-BSS algorithm also works in noisy conditions. However, the presence of background noise expectedly degrades the interferer signal suppression.

6.3.2 Experiments with the Two-Element Distributed Array Mounted on the Car Ceiling

This section presents the separation results that are obtained with the two-element distributed array mounted on the car ceiling, which is depicted in Fig. A.3. We used the same source signals as for the four-element compact array mounted in the rear-view mirror.

Batch Algorithms

As in Sect. 6.3.1, we apply batch algorithms to compare the self-closed algorithm (6.51) with the non-self-closed algorithm (6.53). Figure 6.8 shows the results. They may be compared to those obtained in Sect. 6.3.1 with the two outermost microphones (Fig. 6.4, lower plot). As can be seen in Fig. 6.8, the initial convergence is slightly faster with the two-element distributed array mounted on the car ceiling. However, after $n = 100$ iterations, the SIR improvement attains 12 dB for both setups. Again, the self-closed update leads to the best results.

Online Algorithms

The performance of block-wise and sample-wise online NG-SOS-BSS algorithms is examined. For the block-wise online algorithms, the same settings as in Sect. 6.3.1 are used ($K = \beta = N_{\text{iter}} = 1$). The results are shown in Fig. 6.9. The following comments can be made.

- In terms of SIR improvement, the online separation performance attains 20 dB. This is comparable to the results obtained with the four microphones of the four-element compact array mounted in the rear-view mirror.

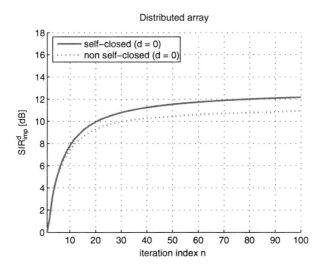

Fig. 6.8. Comparison of the self-closed algorithm (6.51) and non-self-closed algorithm (6.53) in batch mode for the two-element distributed array mounted on the car ceiling. The delay d is set to $d = 0$

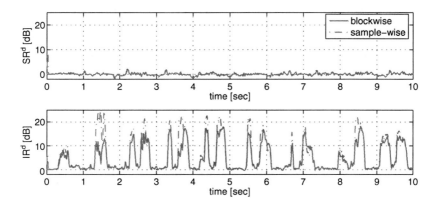

Fig. 6.9. Comparison of block-wise and sample-wise online NG-SOS-BSS algorithms for the two-element distributed array mounted on the car ceiling. The self-closed block-wise algorithm (6.51) has parameters $K = \beta = N_{\text{iter}} = 1$. The sample-wise algorithm is given by (6.67). In both cases, the delay d is set to $d = 0$

- The sample-by-sample algorithm (6.67) convergences significantly faster than the block-online algorithm. Its peak performance is also superior. Both algorithms do not cause any noticeable reduction of the target signal level.

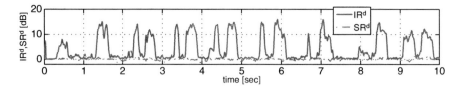

Fig. 6.10. Performance of the sample-wise NG-SOS-BSS algorithm (6.67) with the two-element distributed array mounted on the car ceiling and background noise recorded at $100\,\mathrm{km\,h^{-1}}$, the signal-to-noise ratio at the microphones amounts to 10–15 dB. The delay d is set to $d = 0$

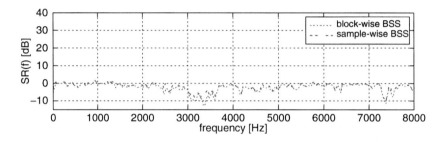

Fig. 6.11. Reduction of the desired signal level as a function of frequency for NG-SOS-BSS algorithms with the distributed array (no background noise)

We applied the sample-wise algorithm to noisy microphone recordings. The result, given in Fig. 6.10, shows that the NG-SOS-BSS algorithm (6.67) also works in noisy conditions with this microphone setup.

The reduction of the desired signal level as a function of the frequency is illustrated in Fig. 6.11. The block-wise and sample-wise algorithms seem to cause little distortion of the desired signal. The distortion caused by NG-SOS-BSS algorithms seems similar to that of the DTD-NLMS algorithm, as may be seen by comparing Fig. 6.11 with Fig. 4.10.

6.3.3 Comparison with Other BSS Algorithms in the Frequency Domain

This section compares the time-domain online NG-SOS-BSS block-wise batch algorithm with other widely used online frequency-domain BSS algorithms. First, we present briefly how BSS can be applied in the frequency domain. (For more details on the frequency-domain BSS algorithms, we refer to [73, 76].) Starting with (5.14) on page 67,

$$\mathbf{y}(p) = \mathbf{W}\mathbf{x}(p), \tag{5.14}$$

we derive the corresponding relationship in the frequency domain by applying the short-time discrete Fourier transform (STFT):

$$\mathbf{Y}^{(f)}(k) = \mathbf{W}^{(f)}\mathbf{X}^{(f)}(k). \tag{6.70}$$

Table 6.5. Parameters for the transformation to frequency domain

parameter	value
window	Hanning
frame length	512
frame shift	256
FFT length	1024

Here, the complex vectors $\mathbf{Y}^{(f)}(k)$ and $\mathbf{X}^{(f)}(k)$ contain the fth frequency bin of the signals $\mathbf{y}(p)$ and $\mathbf{x}(p)$ in the kth time frame. The vectors $\mathbf{Y}^{(f)}(k)$ and $\mathbf{X}^{(f)}(k)$, and the matrix $\mathbf{W}^{(f)}$ have sizes $N \times 1$, $M \times 1$, and $N \times M$, respectively. The analysis parameters are summarized in Table 6.5.

Let us point out that (5.14) and (6.70) are not equivalent, since the latter implies circular convolution. In (6.70), $\mathbf{W}^{(f)}$ separates the sources for each frequency bin f in an instantaneous manner. This implies a narrowband signal model where the mixing process can be written in the frequency domain as

$$\mathbf{X}^{(f)}(k) = \mathbf{H}^{(f)}\mathbf{S}^{(f)}(k), \tag{6.71}$$

by applying the short-time discrete Fourier transform to (5.15). In (6.71), it is assumed that the mixing process is performed in each frequency bin independently.

We compare the time-domain BSS algorithm in (6.51) with two widely used frequency-domain algorithms, the first by Parra et al. [76] and the second by Mukai et al. [73]. The first algorithm, similar to the time-domain algorithm, is based on second-order statistics. The separation matrix $\mathbf{W}^{(f)}$ is adjusted to minimize the least-square cost function

$$J_{\mathrm{BSS,LS}} \triangleq \sum_k \|\mathbf{R}_{\mathbf{yy}}^{(f)}(k) - \mathrm{diag}\left(\mathbf{R}_{\mathbf{yy}}^{(f)}(k)\right)\|^2. \tag{6.72}$$

The $N \times N$ matrix $\mathbf{R}_{\mathbf{yy}}^{(f)}$ denotes the output cross correlation at the frequency f and the norm is defined as $\|\mathbf{R}\|^2 = \mathrm{tr}\left(\mathbf{R}^H\mathbf{R}\right)$. The cost function $J_{\mathrm{BSS,LS}}$ is a diagonalization criterion on the output correlation matrix and is in this respect similar to the cost function in (5.34). This block-wise batch algorithm is performed on the current data block ($K = 1$) with one iteration ($N_{\mathrm{iter}} = 1$).

The second frequency-domain algorithm exploits the higher-order statistics (HOS). Higher-order statistics are more difficult to estimate than the second-order statistics. Hence, a large batch size is used ($K = 20$). The separation matrix $\mathbf{W}^{(f)}$ is adjusted iteratively using the Infomax algorithm [13] and the natural gradient, as in [9]. The update may be written as[10]

$$\Delta\mathbf{W} = -\mu\left(\widehat{\mathbf{E}}\left\{\Phi(\mathbf{Y})\mathbf{Y}^H\right\} - \mathrm{diag}\,\widehat{\mathbf{E}}\left\{\Phi(\mathbf{Y})\mathbf{Y}^H\right\}\right)\mathbf{W}, \tag{6.73}$$

---

[10] For brevity, we omit the frequency argument f and the time frame argument k. The operator $\widehat{\mathbf{E}}\{\}$ denotes the averaging over K time frames.

where Φ is a nonlinear function defined as $\Phi(\mathbf{Y}) = \tan h(g|\mathbf{Y}|) \exp(-i\varphi(\mathbf{Y}))$ and $\varphi(\mathbf{Y})$ is the phase of $\mathbf{Y}$. After a batch of $\beta L = 20L$ new samples has been received, the algorithm (6.73) runs with $N_{\text{iter}} = 4$ iterations. The parameters β and N_{iter} are set so that the algorithms in question require similar amounts of computation. For experiments with the four-element compact array mounted in the rear-view mirror, this natural gradient algorithm was applied with PBSS. To this end, we simply replace the operation $\text{diag}(\cdot)$ by $\text{diag}_{1,[2,3,4]}(\cdot)$ in (6.73), as described in Sect. 5.4.

Frequency-domain BSS has the drawback that different permutations may occur in different frequency bands in an inconsistent manner. This problem is solved by initializing the separation system with (6.69) for $d = 0$ so that the input signals are not delayed. This removes the permutation ambiguity at the beginning of the adaptation, as described in Sect. 6.1.6. Moreover, we constrain the separation filters to length $L = 256$ in the time domain, i.e., a fourth of the transformation length [20]. This constraint has a smoothing effect on the separation filters, which helps to prevent the permutation [75]. It also avoids circularity effects of the convolution in the frequency domain. In our experiments, these countermeasures were sufficient to avoid permutation inconsistencies.

It is not always easy to compare different adaptive algorithms fairly. In particular, the step-size parameter may significantly influence the separation performance. To obtain an objective performance measure, the quantities $Q_{[0,3]}$ and $Q_{[3,10]}$ defined in Table 2.1 are considered. The value $Q_{[0,3]}$ is used as an approximate measure of the speed of convergence during the initial convergence phase. $Q_{[3,10]}$ gives an approximate measure of the performance after the initial convergence.

By adjusting the step-size parameter for each algorithm in question, their speed of convergence may be made approximately equal. In other words, we use $Q_{[0,3]}$ as a reference to calibrate the step-sizes. Then, the performance indexes $Q_{[3,10]}$ may be comparable to each other. (This approach is not the only possible one. For completeness, it would be fair for example to also tune the step-size parameter for each algorithm to a certain final $Q_{[3,10]}$ and then measure the initial convergence $Q_{[0,3]}$.)

The results of the three algorithms are compared in Table 6.6. In the case of the time-domain NG-SOS-BSS algorithm, the self-closed update is employed

Table 6.6. Comparison of the presented time-domain NG-SOS-BSS algorithm with frequency-domain BSS algorithms (*FD* frequency domain, *HOS* high-order-statistics, *SOS* second-order-statistics)

algorithm	compact array		distributed array	
	$Q_{[0,3]}$ (dB)	$Q_{[3,10]}$ (dB)	$Q_{[0,3]}^{\text{d}}$ (dB)	$Q_{[3,10]}^{\text{d}}$ (dB)
NG-SOS-BSS (6.54),(6.51)	4.0	10.2	6.7	11.4
FD-SOS [76]	4.1	6.9	6.7	8.6
FD-HOS [73]	4.1	7.8	6.8	10.0

with $d = 0$ and the block-wise adaptation parameters $K = \beta = N_{\text{iter}} = 1$. We can see that the performance $Q_{[3,10]}$ is limited for the frequency-domain algorithms. This may be explained by the narrowband assumption in the mixing model (6.71). The model (6.71) is valid for narrowband source signals, which is unrealistic for speech sources and reduces the performance. The FD-SOS algorithm has the worst performance due to the inferior statistical efficiency of the cost function $J_{\text{BSS,LS}}$ (see Sect. 8.1.2).

6.4 Summary and Conclusion

The first part of this chapter proposed a rigorous treatment of the redundancy issue that is related to the Sylvester matrices. The problem was tackled by deriving general convolutive formulations of the natural gradient for an arbitrary cost function J. Two types of updates arise from this analysis: the self-closed updates, which depend on a parameter d controlling the acausal length (or delay) in the separation system, and the non-self-closed updates. It emerged that any row or the Lth column may be chosen as reference to maintain the Sylvester structure, in addition to the two choices given in [5]. Our derivation provided self-closed and non-self-closed update rules for both causal and acausal separation systems. In particular, self-closed update rules for acausal separation systems have been obtained.

It appeared that causal mixing/separation systems have a special status: In these systems a group structure can be defined using the self-closed operation $\star_0$. Moreover, using causal separation systems removes the permutation ambiguity. These results are valid not only for NG-SOS-BSS algorithms but also for other BSS algorithms that are based on the natural gradient.

In the case of the NG-SOS-BSS algorithms, it was shown that these natural gradient algorithms involve the inversion of output correlation matrices $\mathbf{R_{yy}}(kL)$. To remove this matrix inversion, the updates were approximated in (6.44) and (6.45). The price for the approximation is that certain gradient terms which do not belong to the Sylvester subspace are involved in the computation of the updates.

The second part of this chapter described online implementations of NG-SOS-BSS algorithms for both block-wise and sample-wise updates. In particular, it was shown that constraining the diagonal filters to simple delays of d taps yields a new sample-wise NG-SOS-BSS algorithm (6.68) with rapid tracking capabilities and $\mathcal{O}(L)$ complexity.

Finally, the applicability of the NG-SOS-BSS algorithms was experimentally assessed using both causal and acausal separation systems in noisy conditions. Self-closed NG-SOS-BSS updates seem to be more robust than their non-self-closed counterparts. For this reason, they will be used for further experiments in this book and the non-self-closed updates will no longer be considered. In particular, the proposed self-closed update for acausal separation systems will be necessary in Chap. 9 when combining NG-SOS-BSS with geometric prior information.

At last, the causal self-closed time-domain NG-SOS-BSS algorithms showed the best performance in a comparison with other widely used frequency-domain BSS algorithms.

To summarize, the most important results of this chapter are:

- Self-closed and non-self-closed implementations of NG-SOS-BSS algorithms have been derived for both causal and acausal separation systems. The mapping from their Sylvester matrices formulation to their actual implementation involves certain approximations, which we have made explicit.
- Setting a causality constraint on the separation system removes the permutation ambiguity. Moreover, *not* setting this constraint prevents equivariant separation performances. However, the causality constraint limits the set of allowed scenarios, as depicted in Fig. 6.2.
- Constraining the diagonal filters to simple delays of d taps yields the new sample-wise NG-SOS-BSS algorithm (6.68) with rapid tracking capabilities and $\mathcal{O}(L)$ complexity.
- We experimentally found out that the self-closed NG-SOS-BSS algorithms (6.51) and (6.54) are more robust and lead to better performance than their non-self-closed counterparts. They also showed a better performance than other widely used frequency-domain algorithms.

7

On the Convergence and Stability
in Second-Order Statistics BSS

In Chap. 6, we have derived natural gradient SOS-BSS algorithms (NG-SOS-BSS) to adapt the separation matrix $\mathbf{W}$. This chapter addresses the question of the convergence of these algorithms, that is, the convergence of $\mathbf{W}$ to a *separating* matrix. An examination of the global convergence is of practical interest because it may indicate the range of admissible step-sizes μ for which these NG-SOS-BSS algorithms globally converge. This issue is discussed in Sect. 7.1, which explains why the analysis of the global convergence is hardly tractable for NG-SOS-BSS algorithms, even in the case of instantaneous mixing. However, in the instantaneous case, an analysis is possible for a simplified decorrelation algorithm that is closely related to NG-SOS-BSS. Then, we propose an interpretation of the results of this analysis for convolutive NG-SOS-BSS. In contrast to the global behavior, the local behavior of convolutive NG-SOS-BSS may be analyzed. This is discussed in Sect. 7.2 where sufficient local stability conditions are given.

7.1 Global Convergence

A necessary condition for the global convergence of an iterative minimization algorithm to a desired solution is that this desired solution is a global minimum of the cost function. This condition is satisfied for the SOS-BSS cost function in (5.34): Its minima correspond to separating matrices $\mathbf{W}$ (i.e., so that $\mathrm{boff}\,(\mathbf{WH}) = \mathbf{0}$ up to source permutations) [20]. However, this necessary convergence condition is not sufficient. Under certain restrictions, the convergence of the gradient descent toward the global minimum of J may be guaranteed. These restrictions are typically placed on the initial point $\mathbf{W}(0)$, on the shape of J (ellipticity, convexity), and on the step-size $\mu(n)$. Unfortunately, the separation cost function J in (5.34) exhibits a rather complex shape, as may be seen in Fig. 5.2. This makes it difficult to guarantee the global convergence of NG-SOS-BSS algorithms.

A result about the global convergence has been already presented in Sect. 6.1.6 with the equivariance property. In the causal case (that is, for $d = 0$ in Chap. 6), the path of the global system

$$\{\overline{\mathbf{C}}(z, n) \text{ for } n = 0, 1, \ldots\}$$

depends on the mixing matrix $\overline{\mathbf{H}}(z)$ only through the initial point $\overline{\mathbf{C}}(z, 0)$. Even though the equivariance property is an indication on the shape of the convergence path, it does not provide any guarantee that a perfect separating solution will be attained for $n \to +\infty$: In other words the off-diagonal elements of the global system, boff$\overline{\mathbf{C}}(z, n)$, may not tend to zeros and we may have[1]

$$\lim_{n \to +\infty} \|\text{boff}\overline{\mathbf{C}}(z, n)\| > 0. \tag{7.1}$$

As explained in [56], global convergence may be obtained with a decreasing step-size $\mu(n)$, so that

$$\sum_n \mu(n) = +\infty \text{ and } \sum_n \mu^2(n) < +\infty. \tag{7.2}$$

Choosing $\mu(n)$ with the property of (7.2) ensures convergence toward a locally stable equilibrium[2] point. Unfortunately, this solution is limited to stationary acoustic environments, since the conditions (7.2) imply that $\mu(n)$ must tend toward zero. This is not applicable to time-varying acoustic environments. In practice, most gradient descent algorithms must be performed with a constant step-size $\mu(n) = \mu$. In this case, the global convergence of the NG-SOS-BSS algorithms is an unsolved problem, even in the "simple" situation of instantaneous mixtures [23]. In particular, we have no upper bound on μ to guarantee the global stability.

7.1.1 Difficulty of a Global Convergence Analysis

To understand the difficulty of a global convergence analysis, it is useful to consider the case of instantaneous mixtures. Let us set $L = L_{\mathrm{m}} = 1$. In this case, the NG-SOS-BSS algorithm (5.41) may directly be applied.[3] If we

---

[1] In (7.1), we consider the convergence to nonpermuted separating solutions. More generally, to take the permuted separating solutions also into account, we should consider the convergence of all elements of $\overline{\mathbf{C}}(z, n)$ except one element per row and column toward zero.

[2] The concept of equilibrium point is defined in Sect. 7.2.

[3] For $L = 1$, the Sylvester constraint "falls": The operator S in (6.9) and its approximations S_d and S^L reduce to the identity,

$$S(\mathbf{A}) = S_d(\mathbf{A}) = S^L(\mathbf{A}) = \mathbf{A} \tag{7.3}$$

for all separation $N \times M$ matrix $\mathbf{A}$. As a consequence, the self-closed and non-self-closed updates (6.51) and (6.53) become identical to (5.40).

consider the evolution of $\mathbf{W}(n)$ as a function of the iteration index n in the gradient descent on a batch of K data blocks, we can write

$$\mathbf{W}(n+1) = \mathbf{W}(n) - \mu \sum_{k=1}^{K} \text{diag}^{-1}\widehat{\mathbf{R}}_{\mathbf{yy}}(k)\text{boff}\left(\widehat{\mathbf{R}}_{\mathbf{yy}}(k)\right)\mathbf{W}(n), \quad (7.4)$$

where $\widehat{\mathbf{R}}_{\mathbf{yy}}(k)$ denotes the estimated output correlation matrix for the kth data block. Without loss of generality, (7.4) can be rewritten as

$$\mathbf{W}(n+1) = f(\mathbf{W}(n)), \quad (7.5)$$

where f is a matrix function which represents the right side of (7.4), and which depends on the input data and on the step-size μ. Two features of f may hamper the global convergence analysis, namely, the nonlinearity and the nonreducibility to dimension one:

(i) Convergence of (7.5) is clearly achieved at a fixed point[4] of f. However, (7.5) may also reach a fixed point of the Cth composition $f^{(C)}$ defined recursively by

$$f^{(C)}(\mathbf{W}) \triangleq f\left(f^{(C-1)}(\mathbf{W})\right) \quad (7.6)$$

$$\text{and } f^{(0)}(\mathbf{W}) = \mathbf{W} \quad (7.7)$$

for any integer $C > 0$. In this case, $\mathbf{W}(n)$ moves cyclically along the path $\{\mathbf{W}, f(\mathbf{W}), \ldots, f^{(C-1)}(\mathbf{W})\}$. To determine such cyclic paths, an equation of the kind $f^{(C)}(\mathbf{W}) = \mathbf{W}$ must be solved. If f is linear, solving this equation is particularly simple, since any composition $f^{(C)}, C > 0$ is also linear. Otherwise, solving $f^{(C)}(\mathbf{W}) = \mathbf{W}$ is not always possible (algebraically), for example, if $f^{(C)}$ is a polynomial of degree larger than four.[5] In this case, the global convergence behavior cannot be predicted by a theoretical analysis.

(ii) In certain cases, f may be "factorable" in the form

$$f(\mathbf{W}) = \begin{pmatrix} f_{11}(w_{11}) & \cdots & f_{1M}(w_{1M}) \\ \vdots & \ddots & \vdots \\ f_{N1}(w_{N1}) & \cdots & f_{NM}(w_{NM}) \end{pmatrix}, \quad (7.8)$$

where each entry f_{ij} depends only on w_{ij}. In such cases one may study each sequence $w_{ij}(n+1) = f_{ij}(w_{ij}(n))$ independently. Even if f is not directly factorable as in (7.8), the matrix sequence (7.5) may be reformulated as a collection of scalar sequences. Then, the analysis of the

---

[4] For the stability of the fixed point $\mathbf{W}$, a sufficient condition is that $\left\|\frac{\partial}{\partial \mathbf{W}} f(\mathbf{W})\right\| < 1$. In this case the fixed point is said to be "attracting" (see [62] for example).

[5] According to the Abel–Ruffini theorem, there is no general solution in radicals to polynomial equations of degree larger than four.

sequence of matrices $\mathbf{W}(n)$ reduces to the analysis of a sequence of scalar numbers. The study of the global convergence is made significantly easier if such a decomposition into subproblems of dimension one can be found.

According to remarks (i) and (ii), the difficulty in the convergence analysis of (7.4) is twofold. Firstly, the update is nonlinear. Secondly, there is no direct decomposition into subproblems of dimension one and hence the lack of available global convergence results in BSS. By contrast, the convergence of certain instantaneous blind *decorrelation* algorithms has been analyzed [31, 32]. There, the convergence analysis may be carried out in dimension one.

7.1.2 Convergence Analysis for a Simplified Algorithm

In this section, we continue with the batch algorithm (7.4) in the instantaneous case $(L = 1)$. First, we alter the natural gradient algorithm (5.40) to a simpler decorrelation algorithm so that global convergence results may be obtained. Next, we revisit the convergence analysis of a blind decorrelation algorithm given in [32]. In particular, we obtain an upper bound on the stepsize that ensures global stability (for the instantaneous, altered algorithm). Then, we attempt to translate these results for the original algorithm (5.41) in the general convolutive case.

Let us consider (7.4) for $K = 1$. Since only one time segment is considered, the nonstationarity of the source signals is not exploited anymore. We omit the block argument k and denote the output correlation matrix at the nth iteration by $\widehat{\mathbf{R}}_{\mathbf{yy}}^{(n)}$. Then, (7.4) can be written as

$$\mathbf{W}(n + 1) = \mathbf{W}(n) - \mu(n)\mathrm{diag}^{-1}\left(\widehat{\mathbf{R}}_{\mathbf{yy}}^{(n)}\right)\left(\widehat{\mathbf{R}}_{\mathbf{yy}}^{(n)} - \mathrm{diag}\left(\widehat{\mathbf{R}}_{\mathbf{yy}}^{(n)}\right)\right)\mathbf{W}(n). \quad (7.9)$$

In [32], the matrix $\mathrm{diag}\left(\widehat{\mathbf{R}}_{\mathbf{yy}}^{(n)}\right)$ in (7.9) is replaced by the identity matrix $\mathbf{I}$, which yields

$$\mathbf{W}(n + 1) = \left(\mathbf{I} - \mu(n)\left(\widehat{\mathbf{R}}_{\mathbf{yy}}^{(n)} - \mathbf{I}\right)\right)\mathbf{W}(n). \quad (7.10)$$

We obtain an algorithm that merely decorrelates the input signals spatially since the sequence $\widehat{\mathbf{R}}_{\mathbf{yy}}^{(n)}$ in (7.10) should converge to $\mathbf{I}$. As we have seen in Sect. 5.2.1, this generally does not achieve the separation of the source signals. Also, by transforming (7.9) into (7.10), we have lost the desired nonholonomicity property expressed in (5.46). Thus, the range of admissible step-sizes $\mu(n)$ depends on the scale of $\widehat{\mathbf{R}}_{\mathbf{yy}}^{(n)}$.

In the following, we determine the range of admissible step-sizes $\mu(n)$ (for each n) so that the sequence $\widehat{\mathbf{R}}_{\mathbf{yy}}^{(n)}$ in (7.10) converges to $\mathbf{I}$. Substituting $\mathbf{W}(n + 1)$ in (7.10) into $\widehat{\mathbf{R}}_{\mathbf{yy}}^{(n+1)} = \mathbf{W}(n + 1)\widehat{\mathbf{R}}_{\mathbf{xx}}\mathbf{W}^{\mathrm{T}}(n + 1)$, we may write $\widehat{\mathbf{R}}_{\mathbf{yy}}^{(n+1)}$ in terms of $\widehat{\mathbf{R}}_{\mathbf{yy}}^{(n)}$ as

$$\widehat{\mathbf{R}}_{\mathbf{yy}}^{(n+1)} = \left(\mathbf{I} - \mu(n)\left(\widehat{\mathbf{R}}_{\mathbf{yy}}^{(n)} - \mathbf{I}\right)\right)\widehat{\mathbf{R}}_{\mathbf{yy}}^{(n)}\left(\mathbf{I} - \mu(n)\left(\widehat{\mathbf{R}}_{\mathbf{yy}}^{(n)} - \mathbf{I}\right)\right)^{\mathrm{T}}. \quad (7.11)$$

Let us denote by f_R the function defined by $\widehat{\mathbf{R}}_{\mathbf{yy}}^{(n+1)} = f_R\left(\widehat{\mathbf{R}}_{\mathbf{yy}}^{(n)}\right)$ in (7.11). Since f_R is a matrix polynomial,[6] $\widehat{\mathbf{R}}_{\mathbf{yy}}^{(n+1)}$ has the same eigenvectors as $\widehat{\mathbf{R}}_{\mathbf{yy}}^{(n)}$. Therefore, the sequence of the output correlations $\widehat{\mathbf{R}}_{\mathbf{yy}}^{(n)}$, for $n = 1, 2, \dots$ may be described by each of its eigenvalues $\lambda_i^{(n)}$, $i = 1, \dots, N$. According to (7.11), we have

$$\lambda_i^{(n+1)} = f_\lambda\left(\lambda_i^{(n)}\right), \quad (7.12)$$

where the function f_λ is defined as $f_\lambda(x) \triangleq (1 - \mu(x-1))^2 x$. To analyze the convergence of the sequence $\lambda_i^{(n)}$, $n = 0, 1, \dots$, we need to determine the fixed points of f_λ. The equation $f_\lambda(x) = x$ has two solutions: $x_1 = 1$ and $x_2 = (2 + \mu)/\mu$. Only x_1 corresponds to a stable equilibrium, since $f_\lambda'(x_2) > 1$. The fixed points of the composition $f_\lambda^{(2)}(\cdot) = f_\lambda(f_\lambda(\cdot))$ are difficult to extract because $f_\lambda^{(2)}$ is a polynomial of degree six. In the following, only the convergence of the eigenvalues $\lambda_i^{(n)}$ toward one is considered, and we define the errors $\tilde{\lambda}_i^{(n)} \triangleq \lambda_i^{(n)} - 1$. The convergence of the errors $\tilde{\lambda}_i^{(n)}$ to zero is equivalent to the convergence of $\widehat{\mathbf{R}}_{\mathbf{yy}}^{(n)}$ to $\mathbf{I}$. Equation (7.11) can be rewritten in terms of eigenvalues $\lambda_i^{(n)}$ as

$$\tilde{\lambda}_i^{(n+1)} = \left(\left(1 - \mu(n)\lambda_i^{(n)}\right)^2 - \mu^2(n)\lambda_i^{(n)}\right)\tilde{\lambda}_i^{(n)}. \quad (7.13)$$

Let us define the second-order polynomial $P_\lambda(x) \triangleq (1 - x\lambda)^2 - x^2\lambda$. As may be seen from (7.13), the error $\tilde{\lambda}_i^{(n+1)}$ decreases at iteration step n if and only if $|P_{\lambda_i^{(n)}}(\mu(n))| < 1$. This condition determines the range of admissible step-sizes. For any $\lambda, \mu > 0$, we have the following necessary condition:

$$\mu < \mu_{\max}(\lambda) \Rightarrow |P_\lambda(\mu)| < 1, \quad (7.14)$$

where $\mu_{\max}(\lambda)$ is defined by

$$\mu_{\max}(\lambda) \triangleq \begin{cases} \left(-\lambda + \sqrt{2\lambda - \lambda^2}\right)/\left(\lambda - \lambda^2\right) & \text{for } \lambda \in]0, 1], \\ 2/(\lambda - 1) & \text{for } \lambda > 1. \end{cases} \quad (7.15)$$

Unfortunately, the presence of a square root and quadratic terms in (7.15) makes it inconvenient to develop a practical normalized step-size. We may

---

[6] The fact that f_R is a polynomial is an essential difference to (7.9). One could similarly try to argue that (7.9) has converged when the eigenvalues of $\mathbf{Q}^{(n)} = \text{diag}^{-1}\left(\widehat{\mathbf{R}}_{\mathbf{yy}}^{(n)}\right)\widehat{\mathbf{R}}_{\mathbf{yy}}^{(n)}$ are all equal to one. Unfortunately, $\mathbf{Q}^{(n+1)}$ is not a polynomial function of $\mathbf{Q}^{(n)}$, which prevents diagonalization.

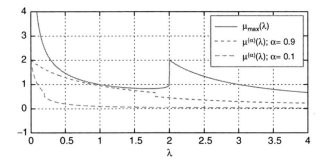

Fig. 7.1. $\mu_{\max}$ in (7.15) as a function of an eigenvalue λ of $\widehat{\mathbf{R}}_{\mathbf{yy}}$ and the practical lower bounds μ_α of $\mu_{\max}$, for $\alpha = 0.9$ and $\alpha = 0.1$ in (7.17)

formulate an assertion that is more conservative than (7.14) if we use a lower bound μ_{lb} of $\mu_{\max}$:

$$\mu < \mu_{\mathrm{lb}}(\lambda) \Rightarrow |P_\lambda(\mu)| < 1. \tag{7.16}$$

In the following we consider a lower bound $\mu_\alpha(\lambda)$ that depends on a free parameter $\alpha \in [0, 1]$ and that is defined as

$$\mu_\alpha(\lambda) \triangleq \begin{cases} 2\alpha/(\alpha + \lambda) & \text{for } 0 < \lambda < 2\alpha, \\ \alpha/\lambda & \text{for } \lambda > 2\alpha. \end{cases} \tag{7.17}$$

$\mu_{\max}(\lambda)$ and $\mu_\alpha(\lambda)$ are plotted for $\alpha = 0.1$ and $\alpha = 0.9$ in Fig. 7.1. It illustrates that the condition in (7.16) is more conservative than (7.14). It should be mentioned that alternative, less conservative simplifications on $\mu_{\max}(\lambda)$ may be developed.

The condition $|P_{\lambda_i^{(n)}}(\mu(n))| < 1$ for all $i = 1, \ldots, N$ is satisfied when

$$0 < \mu(n) < \mu_\alpha\left(\lambda_{\max}^{(n)}\right), \tag{7.18}$$

where $\lambda_{\max}^{(n)}$ is the largest eigenvalue of $\widehat{\mathbf{R}}_{\mathbf{yy}}^{(n)}$. The parameter $\alpha \in]0, 1]$ is set by the user. Its value influences which line of (7.17) should be used to ensure the stability of (7.10).

Hereafter we turn (7.10) into a stable normalized algorithm (that is, an algorithm with a normalized step-size that does not depend on the power of the input signals, just like the NLMS algorithm). Consider first the case $\lambda_{\max}^{(n)} > 2\alpha$, which is likely to occur if α is set close to zero. Then a normalized step-size $\tilde{\mu}(n)$ is given by $\tilde{\mu}(n) = \mu(n)\lambda_{\max}^{(n)}$. In practice, it is desirable to avoid the computation of the eigenvalues. Instead, one could employ the upper bound $\lambda_{\max}^{(n)} < \mathrm{tr}(\widehat{\mathbf{R}}_{\mathbf{yy}}^{(n)})$ as an overestimation and set $\tilde{\mu}(n) = \mu(n)\mathrm{tr}(\widehat{\mathbf{R}}_{\mathbf{yy}}^{(n)})$. This is a reasonable estimate for $\tilde{\mu}(n) = \mu(n)\lambda_{\max}^{(n)}$ only if $\widehat{\mathbf{R}}_{\mathbf{yy}}^{(n)}$ has a single dominant eigenvalue. Likewise, the situation $\lambda_{\max}^{(n)} > 2\alpha$ may be detected in practice with $\mathrm{tr}(\widehat{\mathbf{R}}_{\mathbf{yy}}^{(n)}) > 2\alpha$. This yields:

$$\mathbf{W}(n+1) = \mathbf{W}(n) - \tilde{\mu}(n)\mathrm{tr}^{-1}\left(\widehat{\mathbf{R}}_{\mathbf{yy}}^{(n)}\right)\left(\widehat{\mathbf{R}}_{\mathbf{yy}}^{(n)} - \mathbf{I}\right)\mathbf{W}(n), \qquad (7.19)$$

where $\mathrm{tr}^{-1}(\mathbf{A}) = 1/\mathrm{tr}\mathbf{A}$. The stability conditions are given by (7.18) and the first line of (7.17) and may be written independently of the input data as $0 < \tilde{\mu}(n) < \alpha$.

Now, consider the case $\lambda_{\max}^{(n)} \leq 2\alpha$, which is likely to occur if α is set close to one. According to (7.18), the algorithm (7.10) may be made stable using α as a regularization parameter:

$$\mathbf{W}(n+1) = \mathbf{W}(n) - 2\tilde{\mu}(n)\left(\alpha + \mathrm{tr}\left(\widehat{\mathbf{R}}_{\mathbf{yy}}^{(n)}\right)\right)^{-1}\left(\widehat{\mathbf{R}}_{\mathbf{yy}}^{(n)} - \mathbf{I}\right)\mathbf{W}(n). \quad (7.20)$$

Then the stability conditions are given by (7.18) and the second line of (7.17), and may be written as $0 < \tilde{\mu}(n) < \alpha$ as in the first case.

Let us summarize (7.19) and (7.20) as

$$\mathbf{W}(n+1) = \mathbf{W}(n) - \tilde{\mu}(n)\mathrm{tr}^{-1}\left(\widehat{\mathbf{S}}_{\mathbf{yy}}^{(n)}\right)\left(\widehat{\mathbf{R}}_{\mathbf{yy}}^{(n)} - \mathbf{I}\right)\mathbf{W}(n), \qquad (7.21)$$

$$\text{where } \widehat{\mathbf{S}}_{\mathbf{yy}}^{(n)} = \begin{cases} \widehat{\mathbf{R}}_{\mathbf{yy}}^{(n)} & \text{for } \mathrm{tr}(\widehat{\mathbf{R}}_{\mathbf{yy}}^{(n)}) > 2\alpha \\ \frac{1}{2}\left(\alpha + \widehat{\mathbf{R}}_{\mathbf{yy}}^{(n)}\right) & \text{for } \mathrm{tr}(\widehat{\mathbf{R}}_{\mathbf{yy}}^{(n)}) \leq 2\alpha, \end{cases} \qquad (7.22)$$

where $\tilde{\mu}(n)$ should be chosen in the range $0 < \tilde{\mu}(n) < \alpha$.

Interpretation for NG-SOS-BSS

The above analysis has been carried out in the case $L = 1$ for the decorrelation algorithm in (7.10). It has led to the globally stable decorrelation algorithm (7.21) with a normalized step-size. In this paragraph, we infer by similarity a regularization scheme for the instantaneous algorithm (7.9) and for the general convolutive case (5.41) for $L, K > 1$.

Firstly, we observe that (7.21) may be seen as an altered version of (7.9) with the following replacements:

(i) The normalization term $\mathrm{tr}^{-1}\left(\widehat{\mathbf{R}}_{\mathbf{yy}}^{(n)}\right)\mathbf{I}$ in (7.19) is replaced by $\mathrm{diag}^{-1}\left(\widehat{\mathbf{R}}_{\mathbf{yy}}^{(n)}\right)$ in (7.9). In contrast to (7.9), the update in (7.19) is not normalized by an estimate of the power $\sigma_{y_p}^2$ of the output signal y_p but by an estimate of the sum $\sum_n \sigma_{y_n}^2$.

(ii) The gradient direction $\widehat{\mathbf{R}}_{\mathbf{yy}}^{(n)} - \mathbf{I}$ in (7.19) is replaced by $\widehat{\mathbf{R}}_{\mathbf{yy}}^{(n)} - \mathrm{diag}\left(\widehat{\mathbf{R}}_{\mathbf{yy}}^{(n)}\right)$ in (7.9).

Using these two correspondences, the results on the global convergence of (7.21) may be applied to (7.9) as follows:

$$\mathbf{W}(n+1) = \mathbf{W}(n) - \tilde{\mu}(n)\mathrm{diag}^{-1}\left(\widehat{\mathbf{S}}_{\mathbf{yy}}^{(n)}\right)\left(\widehat{\mathbf{R}}_{\mathbf{yy}}^{(n)} - \mathrm{diag}\left(\widehat{\mathbf{R}}_{\mathbf{yy}}^{(n)}\right)\right)\mathbf{W}(n), \quad (7.23)$$

where

$$
\widehat{\mathbf{S}}_{\mathbf{y}_i \mathbf{y}_j}^{(n)} \triangleq
\begin{cases}
\widehat{\mathbf{R}}_{\mathbf{y}_i \mathbf{y}_j}^{(n)} & \text{for } \mathrm{tr}(\widehat{\mathbf{R}}_{\mathbf{y}_i \mathbf{y}_j}^{(n)}) > 2\alpha \\
\frac{1}{2}\left(\alpha + \widehat{\mathbf{R}}_{\mathbf{y}\mathbf{y}}^{(n)}\right) & \text{for } \mathrm{tr}(\widehat{\mathbf{R}}_{\mathbf{y}_i \mathbf{y}_j}^{(n)}) \le 2\alpha
\end{cases}
\quad \text{for } i, j = 1, \dots, N. \quad (7.24)
$$

Note that $\mathrm{tr}(\widehat{\mathbf{R}}_{\mathbf{y}_i \mathbf{y}_j}^{(n)}) = \widehat{\mathbf{R}}_{\mathbf{y}_i \mathbf{y}_j}^{(n)}$, since the matrix $\widehat{\mathbf{R}}_{\mathbf{y}_i \mathbf{y}_j}^{(n)}$ reduces to a scalar factor in the considered case ($L = 1$). The operator $\mathrm{tr}(\cdot)$ is kept to emphasize the similarity between (7.22) and (7.24). Again, it may be noticed that the filter updates are all normalized by the sum of the output energies in (7.21), whereas they are normalized by their individual output energy in (7.23). It should be mentioned that the correspondences (i) and (ii) are simply inferred by similarity. They do not derive from a rigorous analysis of (7.9). Hence, at this point we have no theoretical guarantee that (7.23) is globally stable.

Now we propose a generalization to the convolutive NG-SOS-BSS algorithm in (5.41) for $L, K > 1$. Let us recall the algorithm (5.41):

$$
\mathbf{W}(n+1) = \mathbf{W}(n) - \mu(n) \sum_{k=1}^{K} \mathrm{diag}^{-1}\widehat{\mathbf{R}}_{\mathbf{y}\mathbf{y}}^{(n)}(kL)\,\mathrm{boff}\left(\widehat{\mathbf{R}}_{\mathbf{y}\mathbf{y}}^{(n)}(kL)\right)\mathbf{W}(n). \quad (5.41)
$$

Since the factor α is homogeneous to a signal power, it should not be compared to $\mathrm{tr}(\widehat{\mathbf{R}}_{\mathbf{y}_i \mathbf{y}_j}^{(n)})$ but to $\mathrm{tr}(\widehat{\mathbf{R}}_{\mathbf{y}_i \mathbf{y}_j}^{(n)})/L$. Then, (7.23) could be generalized to $L, K > 1$ as follows:

$$
\mathbf{W}(n+1) = \mathbf{W}(n) - \mu(n) \sum_{k=1}^{K} \mathrm{diag}^{-1}\widehat{\mathbf{S}}_{\mathbf{y}\mathbf{y}}^{(n)}(kL)\,\mathrm{boff}\left(\widehat{\mathbf{R}}_{\mathbf{y}\mathbf{y}}^{(n)}(kL)\right)\mathbf{W}(n), \quad (7.25)
$$

where

$$
\widehat{\mathbf{S}}_{\mathbf{y}_i \mathbf{y}_j}^{(n)}(kL) =
\begin{cases}
\widehat{\mathbf{R}}_{\mathbf{y}_i \mathbf{y}_j}^{(n)} & \text{for } \mathrm{tr}(\widehat{\mathbf{R}}_{\mathbf{y}_i \mathbf{y}_j}^{(n)}(kL))/L > 2\alpha \\
\frac{1}{2}\left(\alpha + \widehat{\mathbf{R}}_{\mathbf{y}\mathbf{y}}^{(n)}\right) & \text{for } \mathrm{tr}(\widehat{\mathbf{R}}_{\mathbf{y}_i \mathbf{y}_j}^{(n)}(kL))/L \le 2\alpha
\end{cases}
\quad (7.26)
$$

for $i, j = 1, \dots, N$. Let us explain why this regularization scheme might be able to prevent instability: In the NG-SOS-BSS algorithm (5.41), the normalization terms are estimates of the output powers $\sigma_{y_n}^2(kL)$. This power may become very small, for example near convergence if the source signal s_n is zero, which may yield very large updates and instability. A dynamic regularization scheme is provided with (7.26). This scheme is integrated in the convolutive NG-SOS-BSS updates (6.51) and (6.53), and in the Partial BSS (PBSS) updates (6.54) and (6.55).

Again let us mention that there is no guarantee of the stability in the convolutive case (even if $\mu < \alpha$). The condition $\mu \in [0, \alpha]$ is sufficient only for the batch instantaneous decorrelation algorithm (7.21). It may not be sufficient anymore for its altered version (7.23), which is close to the NG-SOS-BSS algorithm (5.41) in the instantaneous case. The condition may become even more severely insufficient with the generalization to the convolutive case where it is not clear how the parameter α depends on the filter length L. In practice, we noticed that instability may be prevented if the step-size is small enough (for example, $\mu = 0.002$). The parameter α should be chosen as small as possible (but larger than μ). Then, the regularization is applied near the convergence and not in the transient phase, which allows for fast convergence in the transient phase.

7.2 Local Stability

The concept of stability has several aspects in adaptive signal processing. Since we use a feedforward (as opposed to feedback) filter architecture, the *structural* BIBO (bounded input bounded output) stability is guaranteed [60]. An iterative algorithm is *globally* stable if it converges to some equilibrium point for any initial condition. The global stability is necessary but not sufficient for the convergence to a desired solution (spurious solutions may exist, such as local minima of the cost function), which has been discussed in Sect. 7.1.

The *local* stability describes the behavior of the algorithm in the vicinity of a particular equilibrium point. Local stability results are easier to obtain than global stability results since the algorithm under consideration may be linearized in the vicinity of a separating matrix. This may provide necessary (yet insufficient) conditions for the convergence to a desired solution. Let us now describe this concept in more detail.

Without loss of generality, the expectation of any BSS update may be written as

$$\mathbf{W}(n+1) = \mathbf{W}(n) - \mu g(\mathbf{W}(n)), \tag{7.27}$$

where g is a matrix function that depends on the mixing $\mathbf{H}$ and on the source signal statistics. The separation matrix $\mathbf{W}_{\text{opt}}$ is a equilibrium point (that is, a separating matrix) if and only if $g(\mathbf{W}_{\text{opt}}) = \mathbf{0}$. Let us consider a small deviation $\varepsilon(n)$ around the equilibrium point $\mathbf{W}_{\text{opt}}$:

$$\mathbf{W}(n) = \mathbf{W}_{\text{opt}} + \varepsilon(n). \tag{7.28}$$

The adaptation in (7.27) can be written in terms of $\varepsilon(n)$ as

$$\varepsilon(n+1) = \varepsilon(n) - \mu g\left(\mathbf{W}_{\text{opt}} + \varepsilon(n)\right). \tag{7.29}$$

The first-order approximation of (7.29) is given by

$$\varepsilon(n+1) = \varepsilon(n) - \mu D_g^{(\mathbf{W}_{\text{opt}})}\left(\varepsilon(n)\right), \tag{7.30}$$

where $D_g^{(\mathbf{W}_{\text{opt}})}\left(\varepsilon(n)\right)$ denotes the derivative of g at point $\mathbf{W}_{\text{opt}}$ and is a linear function of $\varepsilon(n)$. If the sequence $\varepsilon(n)$ in (7.30) converges to zero for some $\mu > 0$, then $\mathbf{W}_{\text{opt}}$ is said to be locally stable. This may be interpreted geometrically as follows: The point $\mathbf{W}_{\text{opt}}$ is locally stable if the adaptation term $g(\mathbf{W}_{\text{opt}} + \boldsymbol{\varepsilon}(n))$ points to $\mathbf{W}_{\text{opt}}$, i.e., if a small deviation from $\mathbf{W}_{\text{opt}}$ is pulled back to the equilibrium. Therefore, the local stability is concerned with the *direction* of the adaptation term but not with its norm (as opposed to the global stability). A necessary and sufficient condition for local stability is that all eigenvalues of the linear mapping $D_g^{(\mathbf{W}_{\text{opt}})}$ have positive real parts.

As opposed to the lack of knowledge about global convergence, several local stability results are available in BSS, for example in [8, 39, 52]. However, these results are derived in the vicinity of the inverse of the mixing system, assuming not only separation but also deconvolution of the source signals. In this section, we examine the stability around the vicinity of an equilibrium point $\mathbf{W}_{\text{opt}}$ that merely separates the sources, as encountered in practice.

The stability analysis is derived in the following framework: We consider the two-source two-sensor scenario ($M = N = 2$) as depicted in Fig. 2.6. Here, we assume that the diagonal mixing channels $\mathbf{H}_{nn}, n = 1, 2$ and separation filters $\mathbf{W}_{nn}, n = 1, 2$ are unit responses:

$$\mathbf{w}_{nn} = \mathbf{w}_{nn} = \boldsymbol{\delta}_d \tag{7.31}$$

for a certain d. As mentioned in Sect. 2.3, this special scenario may be considered as a physical model in the situation where the room acoustics is not very reverberant and where each source s_n is placed close to the microphone x_n for $n = 1, 2$. The constraint on the separation filters limits the number of degrees of freedom and assures the uniqueness of the equilibrium point $\mathbf{W}_{\text{opt}}$. For $\mathbf{W} = \mathbf{W}_{\text{opt}}$, the source signals are separated but not deconvolved. Besides, the local stability of BSS algorithms generally depends on the statistics of the source signals, which makes it difficult to interpret them physically. An interesting case occurs when the source signals exhibit periods of silences, as is the case for speech signals. Under these conditions, the local stability conditions may be formulated quite simply. For the sake of readability, the computation of $D_g^{(\mathbf{W}_{\text{opt}})}$ and the subsequent derivations are postponed to Appendix D.

Let us define $H_{mn}(k)$ for $k = 0, \ldots, 2L - 2$ as

$$H_{mn}(k) \triangleq \sum_{p=0}^{2L-2} h_{mn,p}\, e^{2i\pi pk/(2L-1)} \qquad \text{for } n, m = 1, 2 \text{ and } n \neq m. \tag{7.32}$$

$H_{mn}(k)$ is similar to the kth frequency bin of the DFT of h_{mn} padded with $L - 1$ zeros (except for a missing minus sign in the exponent of the transform kernel). Under certain conditions (more details can be found in Appendix D), one may derive a *sufficient* local stability condition:

$$|H_{12}(k)H_{21}(k)| < 1 \qquad \forall k = 0, \ldots, 2L - 2. \tag{7.33}$$

Condition (7.33) sets an upper bound on the amount of cross-talk. Interestingly, a closely related stability condition was found experimentally by

Van Gerven [39]. Let us define $H(k) \triangleq H_{12}(k)H_{21}(k)$, which is the DFT of $h = h_{12} * h_{21}$. Van Gerven has found the following stability condition:

$$\sum_{p=0}^{2L-2} |h_p| < 1. \tag{7.34}$$

Condition (7.34) is more conservative than condition (7.33), since

$$\sum_{p=0}^{2L-2} |h_p| < 1 \Rightarrow |H(k)| < 1 \Leftrightarrow |H_{12}(k)H_{21}(k)| < 1. \tag{7.35}$$

Special case $d = 0$

In the special case of causal mixings of white source signals, i.e., for $d = 0$ in (7.31) and $\mathbf{R}_{s_n s_n} \propto \mathbf{I}$, a finer analysis may be carried out. (See Appendix D for more details.) Here, we can provide the following *necessary and sufficient* stability condition:

$$h_{12,0}h_{21,0} < 1. \tag{7.36}$$

Condition (7.36) is satisfied if $h_{12,0}h_{21,0} = 0$, i.e., if one of the cross channels is strictly causal. Remarkably, the same local stability condition has been derived by Charkani and Deville for causal mixings of white source signals [26]. Note that they considered an equilibrium point where the source signals are not only separated but also deconvolved. By contrast, (7.36) is a stability condition in the vicinity of an equilibrium point where the source signals are only separated. While it is not clear how the condition (7.36) relates to than (7.33), it appears that the condition (7.36) is much weaker than (7.34). Hence, causal mixings seem to be favorable for the local stability.

7.3 Summary and Conclusion

The global convergence of NG-SOS-BSS algorithms still remains an open problem, as for many other SOS-BSS algorithms [23]. There exists no description of the transient behavior of NG-SOS-BSS algorithms, even for instantaneous mixtures. We have explained the reasons for this lack of knowledge. Nevertheless, we could analyze a similar blind decorrelation algorithm. Motivated by this analysis, we proposed a dynamic regularization scheme for BSS. In practice, however, the adaptation parameters should be carefully chosen based on simulated or on real recordings. Note that the choice of these parameters may also depend on the filter length L.

Nevertheless, local stability conditions have been derived in a two-source two-sensor scenario. It appears that the local stability of NG-SOS-BSS depends on the amount of cross-talk. Strict causality of one of the cross channels is a sufficient local stability condition. Hence causal systems seem to be a favorable case, like in Sect. 6.1.6.

The most important results of this chapter are the following:

- The global convergence and stability of NG-SOS-BSS algorithms are not guaranteed. The convergence analysis for an instantaneous decorrelation algorithm has been carried out. This analysis motivated a dynamic regularization scheme for second-order convolutive BSS. Still, appropriate adaptation parameters should be chosen from experiments.
- Necessary local stability conditions set an upper bound on the amount of cross-talk.

Comparison of LCMV Beamforming and Second-Order Statistics BSS

Adaptive LCMV beamforming algorithms have been presented in Chap. 4. In Chaps. 5–7, we have examined NG-SOS-BSS algorithms. Both approaches pursue the same objective, namely to reduce interference signals. This chapter gives a side-by-side comparison of these two approaches.

Originally, convolutive blind signal separation (BSS) techniques have been developed within Widrow's adaptive identification framework [38, 90]. This forms a basis for a natural connection between blind source separation and beamforming. Moreover, as pointed out by Cardoso and Souloumiac in the context of narrowband array processing, BSS techniques achieve the separation by filtering the microphone signals spatially, hence the term "blind beamforming" [21]. This similarity is important to understand the theory. By extracting independent signals out of a mixture, BSS actually forms multiple null beams in the direction of interfering sources [10]. This "equivalence" clearly indicates that sources which are spatially close to each other (or aligned) are *not* any better separated by BSS-based array processing than by adaptive LCMV beamforming.

On the other hand, there are obvious differences between adaptive LCMV beamforming and BSS. For example, time-domain BSS algorithms generally do not require prior information on the sensor arrangement.[1] BSS algorithms do not necessitate a target-free interference reference, since the mixing model in (5.15) takes the target leakage into account. By contrast, LCMV beamformers require a double-talk detector to overcome the target leakage problem.

This chapter is organized as follows: In the first part we will compare theoretical properties of the two approaches. Section 8.1 focuses on the properties of their respective cost functions. The second part is a comparison from the practical standpoint. Section 8.2 compares NG-SOS-BSS algorithms and

---

[1] This is not entirely true in the case of frequency-domain BSS algorithms that overcome the permutation problem by exploiting the source directions of arrival, as in [82, 83].

LMS-adapted LCMV beamforming in terms of complexity. Section 8.4 provides an experimental comparison of their performance-to-complexity ratio.

8.1 Properties of the Cost Functions

The cost functions in LCMV beamforming and SOS-BSS exhibit different shapes, which implies different properties for the convergence of a gradient descent. These differences are stressed in Sect. 8.1.1. Section 8.1.2 examines the accuracy of the optimization with a finite number of samples.

8.1.1 Shape of the Cost Function

For simplicity, we examine the gradient descent of the different cost functions for a given batch of T input data samples. Let us recall the two cost functions in this context:

$$J_{\mathrm{BSS}}(\mathbf{W}) \triangleq \sum_{k=1}^{K} \log \det \mathrm{bdiag}(\widehat{\mathbf{R}}_{\mathbf{yy}}(kL)) - \log \det \widehat{\mathbf{R}}_{\mathbf{yy}}(kL), \qquad (8.1)$$

$$J_{\mathrm{LS}}(\mathbf{w}) \triangleq \frac{1}{T} \sum_{p=1}^{T} y^2(p), \quad \mathbf{w} \text{ s.t. } \mathbf{C}^{\mathrm{T}}\mathbf{w} = \mathbf{c}. \qquad (8.2)$$

In (8.1), $\widehat{\mathbf{R}}_{\mathbf{yy}}(kL)$ represents a generic estimate of the output correlation matrix around time $p = kL$.

Shape and gradient descent for J_{BSS}

In contrast to J_{LS} in (8.2), $J_{\mathrm{BSS}}(\mathbf{W})$ has a continuum of minima, since $J_{\mathrm{BSS}}(\alpha\mathbf{W}) = J_{\mathrm{BSS}}(\mathbf{W})$ for any nonzero scalar α. In fact, the set of the separating matrices $\mathbf{W}$ that minimize J_{BSS} is even larger and coincides with the output permutation and filtering ambiguities. As described in Sect. 6.1.6, setting a causality constraint on the separation matrix (which should be done only for certain source-microphone arrangements) limits the search space a priori and removes the minima that corresponds to permuted solutions. Hence, the performance of SOS-BSS algorithms may depend on this causality constraint. Other properties of the gradient descent for the cost function J_{BSS} have been discussed in Chap. 7. They are summarized in Table 8.1, which compares them to the properties of J_{LS}.

Shape and gradient descent for J_{LS}

The constraint $\mathbf{C}^{\mathrm{T}}\mathbf{w} = \mathbf{c}$ may be implemented using the GSC structure and the decomposition $\mathbf{w} = \mathbf{w}_0 + \mathbf{B}\mathbf{a}$ described in Sect. 3.2. Since only $\mathbf{a}$ is adapted, we may consider J_{LS} as a function of $\mathbf{a}$, $J_{\mathrm{LS}}(\mathbf{a})$. Define the correlation matrix $\widehat{\mathbf{R}}_{\mathbf{x}_B\mathbf{x}_B}$ and the cross-correlation vector $\widehat{\mathbf{r}}_{x_0\mathbf{x}_B}$ as

Table 8.1. Comparison of the SOS-BSS cost function (8.1) and LS cost function (8.2) used in LCMV beamformers: shape of the cost functions and of the gradient descents on a batch of input sample. In the case of BSS, the natural gradient descent is considered. We refer to Chap. 7 for more details on the convergence and stability of BSS

	J_{LS} in (8.2)	J_{BSS} in (8.1)
Cost function	convex	nonconvex (see Fig. 5.2)
shape		
number of optima	1, given by (3.22)	multiple optima (via filtering and permutation)
Gradient descent		(natural gradient)
update	linear	nonlinear
convergence to cyclic paths	no	possible
convergence analysis	exists	unknown
global stability conditions	given in (4.4)	unknown
local stability	yes	under certain conditions (see Sect. 7.2)

$$\widehat{\mathbf{R}}_{\mathbf{x}_B\mathbf{x}_B} \triangleq \frac{1}{T}\sum_{p=1}^{T}\mathbf{x}_B(p)\mathbf{x}_B^{\mathrm{T}}(p), \qquad \widehat{\mathbf{r}}_{x_0\mathbf{x}_B} \triangleq \frac{1}{T}\sum_{p=1}^{T}x_0(p)\mathbf{x}_B^{\mathrm{T}}(p), \qquad (8.3)$$

and assume that $\widehat{\mathbf{R}}_{\mathbf{xx}}$ is invertible. As we have shown in (3.22), the minimum J_{LS} is then unique and is given by a *closed formula*, the Wiener solution:

$$\mathbf{a}_{\mathrm{opt}} = -\widehat{\mathbf{R}}_{\mathbf{x}_B\mathbf{x}_B}^{-1}\widehat{\mathbf{r}}_{x_0\mathbf{x}_B}. \qquad (8.4)$$

To examine how $\mathbf{a}(n)$ converges to $\mathbf{a}_{\mathrm{opt}}$ in the gradient descent, we define the mismatch between the interference canceler and the Wiener solution $\mathbf{a}_{\mathrm{opt}}$ at iteration n as $\mathbf{m}(n) \triangleq \mathbf{a}(n) - \mathbf{a}_{\mathrm{opt}}$. The gradient descent for J_{LS} may now be formulated as $\mathbf{m}(n+1) = f_{\mathrm{LS}}(\mathbf{m}(n))$, where

$$f_{\mathrm{LS}}(\mathbf{m}(n)) \triangleq \left(\mathbf{I} - \mu\widehat{\mathbf{R}}_{\mathbf{x}_B\mathbf{x}_B}\right)\mathbf{m}(n) \qquad (8.5)$$

(see, e.g., [92]). Let us contrast the properties of the function f_{LS} with the properties of the matrix function f for NG-SOS-BSS algorithms mentioned in Sect. 7.1:

- First, the function f_{LS} is linear and its only[2] fixed point is $\mathbf{m} = \mathbf{0}$.

---

[2] The uniqueness of the fixed point may be easily shown. Since $\widehat{\mathbf{R}}_{\mathbf{x}_B\mathbf{x}_B}$ is positive definite, its unique fixed point is $\mathbf{m} = \mathbf{0}$. The question arises whether paths of the kind $\{\mathbf{m}, f_{\mathrm{LS}}(\mathbf{m}), \ldots, f_{\mathrm{LS}}^{(C-1)}(\mathbf{m})\}$ with $f_{\mathrm{LS}}^C(\mathbf{m}) = \mathbf{m}$ exist. For any $C > 1$, the fixed points of the composition $f^{(C)}$ are the solutions of

$$f_{\mathrm{LS}}^{(C)}(\mathbf{m}) = (\mathbf{I} - \mu\widehat{\mathbf{R}}_{\mathbf{x}_B\mathbf{x}_B})^C\mathbf{m} = \mathbf{m}. \qquad (8.6)$$

Since $(\mathbf{I} - \mu\widehat{\mathbf{R}}_{\mathbf{x}_B\mathbf{x}_B})^C$ is also positive definite, the only fixed point of $f_{\mathrm{LS}}^{(C)}$ is $\mathbf{m} = \mathbf{0}$. Consequently, the algorithm can converge only to $\mathbf{m} = \mathbf{0}$. Therefore, no orbit exists.

- Second, f_{LS} may be diagonalized using the eigenvalue decomposition of $\widehat{\mathbf{R}}_{\mathbf{x}_B\mathbf{x}_B}$, the vector equation $\mathbf{m}(n+1) = \widehat{\mathbf{R}}_{\mathbf{x}_B\mathbf{x}_B}\mathbf{m}(n)$ is then decomposed into a collection of scalar equations by diagonalizing $\widehat{\mathbf{R}}_{\mathbf{x}_B\mathbf{x}_B}$ (see, for example, [92] for more details). Thus, the convergence analysis of the gradient descent reduces to the convergence of a sequence of scalar numbers, i.e., it may be carried out in one dimension.

The convergence properties of the gradient descent are well known [45, 91]. Global convergence in the mean to the Wiener solution is guaranteed if the step-size μ satisfies

$$0 < \mu < \frac{1}{\lambda_{\max}}, \tag{8.7}$$

where $\lambda_{\max}$ denotes the largest eigenvalue of $\widehat{\mathbf{R}}_{\mathbf{x}_B\mathbf{x}_B}$. The Wiener solution is also locally stable.

Minima of the cost functions

The question arises as to how the minima of the cost function $J_{BSS}(\mathbf{W})$ in (8.1) are related to the minimum of $J_{LS}(\mathbf{w})$ in (8.2). This question may hardly be answered in the general case since no closed formula for the minimum of $J_{BSS}(\mathbf{W})$ is known yet. Nevertheless, one might ask whether separating solutions (discussed in Sect. 2.3) are minima of the cost functions in $J_{BSS}(\mathbf{W})$ and $J_{LS}(\mathbf{w})$. Some results may be stated under restricting conditions. Let us assume that interference-independent separating solutions exist (which may be the case if the conditions for spatial separation given in Sect. 2.3.1 are fulfilled). Then these separating solutions are minima of $J_{BSS}(\mathbf{W})$ since the output signals are independent.[3] For these separating solutions to be minima of $J_{LS}(\mathbf{w})$, it is additionally required

- that the desired source is silent (as assumed in [10, 11]),
- or that the desired source does not leak into the interference reference.

If none of these conditions is fulfilled, then (3.38) shows that the minimum of $J_{LS}(\mathbf{w})$ may not be a separating solution and depends on the adequation of the linear constraint $\mathbf{C}^T\mathbf{w} = \mathbf{c}$ with the acoustic channels and of the power of the desired source signals relative to the power of the interferences.

8.1.2 On the Estimation Variance

The SOS-BSS and LCMV cost functions, J_{BSS} in (8.1), and J_{LS} in (8.2), have been defined with a batch of T samples. They may be seen as estimates of underlying, statistically defined cost functions ξ_{BSS} and ξ_{LMS} defined as follows

---

[3] This assumes the empirical correlation of the source signals is zero. This aspect is discussed in Sect. 8.1.2.

$$\xi_{\text{BSS}}(\mathbf{W}) \triangleq \sum_{k=1}^{K} \log \det \text{bdiag}(\mathbf{R}_{\mathbf{yy}}(kL)) - \log \det \mathbf{R}_{\mathbf{yy}}(kL), \qquad (8.8)$$

$$\xi_{\text{LMS}}(\mathbf{w}) \triangleq \mathbf{E}\left\{y^2(p)\right\} \quad \mathbf{w} \text{ s.t.} \quad \mathbf{C}^{\text{T}}\mathbf{w} = \mathbf{c}. \qquad (8.9)$$

Similarly, in (8.8), the matrix $\mathbf{R}_{\mathbf{yy}}(kL)$ represents the true output correlation matrix at time $p = kL$. Let us denote the optimal filters for (8.8) and (8.9) by $\mathbf{W}_{\text{opt}}$ and $\mathbf{w}_{\text{opt}}$, respectively. In this section, the optimal filters for the batch cost functions in (8.1) and (8.2) are seen as batch estimates for $\mathbf{W}_{\text{opt}}$ and $\mathbf{w}_{\text{opt}}$, respectively. These estimates are denoted by $\widehat{\mathbf{W}}_{\text{opt}}$ and $\widehat{\mathbf{w}}_{\text{opt}}$, respectively.

Definition of the estimation variance

In general, the accuracy of a certain estimator with a finite sample size T is characterized by its variance. In the following we briefly explain the concepts of estimator and estimator variance (see, e.g., [69] for more details). The input samples in the vector $\mathbf{x}(p)$ are modeled as realizations of a (multidimensional) stochastic process $\mathbf{X}(p)$. Let $\boldsymbol{\theta}$ be a parameter of $\mathbf{X}(p)$, which we want to estimate. In our context, $\boldsymbol{\theta}$ would represent optimal separation filters. Note that $\boldsymbol{\theta}$ is not a random variable. An estimator for $\boldsymbol{\theta}$ is a random variable $\widehat{\boldsymbol{\theta}}$. The estimator is said to be unbiased if and only if $\mathbf{E}\left\{\widehat{\boldsymbol{\theta}}\right\} = \boldsymbol{\theta}$. The covariance matrix for the estimator $\widehat{\boldsymbol{\theta}}$ is defined as

$$\mathbf{C}\left(\widehat{\boldsymbol{\theta}}\right) \triangleq \mathbf{E}\left\{\left(\widehat{\boldsymbol{\theta}} - \boldsymbol{\theta}\right)\left(\widehat{\boldsymbol{\theta}} - \boldsymbol{\theta}\right)^{\text{T}}\right\}. \qquad (8.10)$$

Its variance $V(\widehat{\boldsymbol{\theta}})$ is the norm of the covariance matrix, $V(\widehat{\boldsymbol{\theta}}) \triangleq \|\mathbf{C}(\widehat{\boldsymbol{\theta}})\|$. In the scalar case, $V(\widehat{\theta}) = \mathbf{E}\left\{(\widehat{\theta} - \theta)^2\right\}$. The variance of the estimator provides a measure of the quality of the estimation. In our context, the quantity to estimate is the minimum of a certain cost function. Since we are not interested in the minimization procedure (such as the gradient descent), only the performance *after* convergence is relevant here.

Illustration in the case of LCMV beamforming

Let us illustrate this concept in the case of LCMV beamforming. It is difficult to obtain meaningful and tractable results on the estimator variance in the general case since the estimator variance depends on the source processes (the source signals). Nevertheless, there is a very special case where the estimator variance may be derived easily and that is worth mentioning. Let us examine the case of instantaneous mixtures and assume that the target source $s_1(p)$ is not active. Now, it is not difficult to show that $\mathbf{w}_{\text{opt}} = \widehat{\mathbf{w}}_{\text{opt}}$ for any sample size T: Using the source correlation matrix $\mathbf{R}_{\mathbf{ss}} = \mathbf{E}\left\{\mathbf{s}(p)\mathbf{s}^{\text{T}}(p)\right\}$ and the estimated correlation matrix $\widehat{\mathbf{R}}_{\mathbf{ss}} = \frac{1}{T}\sum_{p=1}^{T}\mathbf{s}(p)\mathbf{s}^{\text{T}}(p)$, we may write the power of the output signal as follows

$$\mathbf{E}\left\{y^2(p)\right\} = \mathbf{w}^\mathrm{T}\mathbf{H}\mathbf{R}_{\mathrm{ss}}\mathbf{H}\mathbf{w}, \tag{8.11}$$

$$\frac{1}{T}\sum_{p=1}^{T} y^2(p) = \mathbf{w}^\mathrm{T}\mathbf{H}\widehat{\mathbf{R}}_{\mathrm{ss}}\mathbf{H}\mathbf{w}. \tag{8.12}$$

Since the source s_1 is not active, a necessary and sufficient condition to solve (8.2) and (8.9) is

$$\mathbf{C}^\mathrm{T}\mathbf{w} = \mathbf{c}, \quad \mathbf{w}^\mathrm{T}\mathbf{H}_{\mathrm{int}} = \mathbf{0}. \tag{8.13}$$

The matrix $\mathbf{H}_{\mathrm{int}}$ is the mixing matrix for the interferers $s_n(p)$, $n = 2, \ldots, N$:

$$\mathbf{H}_{\mathrm{int}} = \begin{pmatrix} h_{12,0} & \cdots & h_{1N,0} \\ \vdots & & \vdots \\ h_{M2,0} & \cdots & h_{MN,0} \end{pmatrix}. \tag{8.14}$$

If $M = N$, (8.13) determines a unique $\mathbf{w}$ which is the solution of both (8.2) and (8.9). This shows that the solutions of (8.2) and (8.9) coincide. In other words, $\mathbf{w}_{\mathrm{opt}}$ may be perfectly estimated with a finite sample size. This phenomenon is sometimes called statistical *superefficiency* and occurs only for noiseless input signals [24].

Illustration in the case of SOS-BSS

As indicated by Cardoso and Pham, BSS also have a potential for superefficiency [24]. For SOS-BSS cost functions J that are joint diagonalization criteria for the output cross-correlation matrices, this potential becomes concrete when, at any time p, only one source is active:

$$\forall p \quad \exists s_n(p) \neq 0 \quad \Rightarrow \quad s_{n'}(p) = 0 \quad \forall n' \neq n. \tag{8.15}$$

Consider the case of instantaneous mixtures for two sources and the separating matrix given by

$$\begin{pmatrix} w_{11,0} & w_{12,0} \\ w_{21,0} & w_{22,0} \end{pmatrix} = \begin{pmatrix} h_{22,0} & -h_{12,0} \\ -h_{21,0} & h_{11,0} \end{pmatrix}. \tag{8.16}$$

At this point, the output $y_n(p)$ is proportional to the source $s_n(p)$, and according to (8.15) we have

$$\forall p \quad \exists y_n(p) \neq 0 \quad \Rightarrow \quad y_{n'}(p) = 0 \quad \forall n' \neq n. \tag{8.17}$$

Hence, for any time p, the *estimated* output cross-correlation matrix $\widehat{\mathbf{R}}_{\mathbf{yy}}(p)$ is diagonal. Then, any cost function J that is a joint diagonalization criterion for the output cross-correlation matrices is minimum in (8.16).

The case of the cost function J_{BSS} in (8.1) is somewhat particular. If $\mathbf{W}$ is set as in (8.16) and if (8.17) holds, the output correlation matrix $\widehat{\mathbf{R}}_{\mathbf{yy}}(p)$ is singular and the "true" cost function (8.8) as well as its estimate (8.1) are not defined. However we may, say these two cost functions (the estimated one (8.1) and the "true" one (8.8)) are both minimal in (8.16) in the sense that their regularized natural gradients vanish. If (8.15) does not hold, then the separation accuracy may be limited by the empirical cross correlation of the source signals and SOS-BSS methods which exploit the nonstationarity of the source signals are, in general, not superefficient [9].

Comparison of (8.1) with another SOS-BSS cost function based on the nonstationarity

Not all BSS cost functions are equivalent with respect to estimation variance. For example, let us consider the following least-square criterion

$$J_{\text{BSS,LS}}(\mathbf{W}) = \sum_{k=1}^{K} \left\| \widehat{\mathbf{R}}_{\mathbf{yy}}(kL) - \text{bdiag} \left(\widehat{\mathbf{R}}_{\mathbf{yy}}(kL) \right) \right\|^2. \tag{8.18}$$

Similarly to the mutual-information-based cost function (8.1), the cost function in (8.18), which was introduced in Sect. 6.3.3, is a joint diagonalization criterion for the output cross-correlation matrices $\widehat{\mathbf{R}}_{\mathbf{yy}}(kL), k = 1, \ldots, K$. This criterion was introduced heuristically by Parra and Spence for frequency-domain BSS [75]. In the following, we numerically examine the estimation error for two cost functions (8.1) and (8.18).

To this end, we consider the two-source two-sensor scenario and set the sample size $T = 100$. The source signal samples are independently drawn from a normal distribution (white Gaussian noise). The source signal $s_2(p)$ for $p = 1, \ldots, T$ has unit energy. We vary the power of the source signal $s_1(p)$ on its first 25 samples, thus varying the input SIR. A randomly selected instantaneous mixing system is applied to these source signals. The instantaneous mixing model is chosen not just for its simplicity. As explained in Sect. 2.3.1, it also ensures that the separation coefficients are independent of the source signals. Only the source signals and the sample size T influence the separation accuracy. The BSS cost functions (8.1) and (8.18) are minimized numerically, which yield the minimum $\widehat{\mathbf{W}}_{\text{opt}}$. Since the mixing matrix $\mathbf{H}$ is known in our simulation, the estimation accuracy may be quantified using the global system $\mathbf{C} = \widehat{\mathbf{W}}_{\text{opt}}\mathbf{H}$. We define the estimation error $e_{\text{estim}} = c_{12}^2/c_{11}^2$ to measure the quality of the separation at the output $y_1(p)$. The error e_{estim} is averaged over 20 trials. The results are shown in Fig. 8.1.

The superefficiency of the mutual-information-based cost function (8.1) manifests itself by the continuously decreasing estimation error for decreasing SIR_{in}. By contrast, for the cost function (8.18), the estimation error e_{estim} has a lower bound (which depends on the sample size). In this respect, the mutual information criterion (8.1) outperforms the heuristic criterion (8.18). This may be explained as follows: *Not* all output cross-correlation matrices

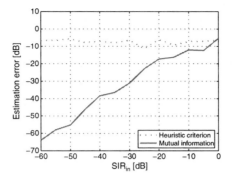

Fig. 8.1. Estimation error $e_{\text{estim}} = \frac{c_{12}^2}{c_{11}^2}$ for the mutual-information-based cost function (8.1) and the heuristic cost function (8.18). The sample size is $T = 100$

$\widehat{\mathbf{R}}_{\mathbf{yy}}(kL), k = 1, \ldots, K$ are equally important in the joint diagonalization criterion (8.1). As may be seen from the natural gradient in (5.41), the mutual information criterion yields a normalization through the output power $\widehat{\mathbf{E}}\left\{y_1^2(kL)\right\}$. This normalization assigns a larger weight to the cross-correlation matrices $\widehat{\mathbf{R}}_{\mathbf{yy}}(kL)$ for which $y_1(p)$ has low power (that is, for $p = 1, \ldots, 25$ in our simulation). Hence, if s_1 is silent in a given interval, an infinite weight is assigned to the cross-correlation matrices $\widehat{\mathbf{R}}_{\mathbf{yy}}(kL)$ in this interval. These matrices are precisely those that allow superefficiency. Therefore, if the target source is silent then the estimation error vanishes. On the other hand, such a weight distribution does not exist with the cost function (8.18). In fact, its gradient

$$\frac{\partial J_{\text{BSS,LS}}}{\partial \mathbf{W}} = \sum_{k=1}^{K} \left(\widehat{\mathbf{R}}_{\mathbf{yy}}(kL) - \text{bdiag}\,\widehat{\mathbf{R}}_{\mathbf{yy}}(kL)\right)\widehat{\mathbf{R}}_{\mathbf{yx}}(kL) \tag{8.19}$$

does not include any normalization. To improve the stability during adaptation, Parra [75] suggests normalizing the learning rule by $\left\|\widehat{\mathbf{R}}_{\mathbf{yx}}(kL)\right\|^2$ as follows:

$$\mathbf{\Delta W} = -\mu \sum_{k=1}^{K} \frac{\partial \left\|\widehat{\mathbf{R}}_{\mathbf{yy}}(kL) - \text{bdiag}\left(\widehat{\mathbf{R}}_{\mathbf{yy}}(kL)\right)\right\|^2}{\partial \mathbf{W}} \Big/ \left\|\widehat{\mathbf{R}}_{\mathbf{yx}}(kL)\right\|^2. \tag{8.20}$$

Unfortunately, this normalization does not seem to allow any useful weight distribution over the matrices $\widehat{\mathbf{R}}_{\mathbf{yy}}(kL), k = 1, \ldots, K$.

8.2 Complexity

This section compares the amount of computation required by LMS-adapted LCMV beamforming and NG-SOS-BSS adaptive algorithms. The complexity of a real multiplication is denoted by C_M and that of a real addition by C_A (divisions are treated as multiplications).

In the following, we calculate the complexity for elementary adaptive "blocks." In the case of the interference canceler, the elementary block is shown in Fig. 8.2a. It includes M' interferer reference signals. By setting $M' = M - 1$, one obtains the complexity for the interference canceler in the GSC. The computational demand for the RGSC may also be evaluated using the complexity of this elementary adaptive block. Setting $M' = 1$ yields the complexity for one filter of the adaptive blocking matrix of the RGSC. We note that the steering delays may require fractional delay filters in the case of the GSC with a compact microphone array [64]. This results in an additional computational cost which depends on the length of the interpolation filters and is not taken into account here. Also, additional computational cost for a control mechanism is not taken into account.

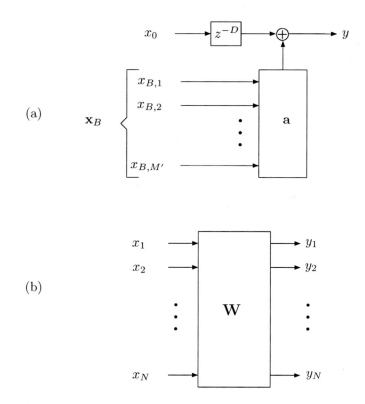

Fig. 8.2. Adaptive elements of the evaluation of the computational complexity: (a) adaptive interference canceler, (b) adaptive SOS-BSS separation block

We distinguish two types of implementation:

- For the *block-wise* batch implementation, the filters are estimated periodically using a batch algorithm after a block of βL new input samples has arrived. The batch algorithm may be iterated N_{iter} times. If we denote the number of operations for one batch iteration by C^{batch}, the corresponding block-wise batch algorithm then has complexity

$$C^{\text{block-wise}} = \frac{N_{\text{iter}}}{\beta L} C^{\text{batch}}. \tag{8.21}$$

The block convolution may be realized efficiently in the DFT domain.
- For *sample-wise* algorithms, the filters are updated after each new input sample as in (4.7) or (6.68).

In the case of SOS-BSS (Fig. 8.2b), the block-wise implementation additionally depends on the number of blocks K as defined in Table 6.2.

Convolution in the DFT domain

As a preliminary, we derive the number of operations for the convolution of two vectors using the fast Fourier transform (FFT). We assume that the calculation of one DFT or inverse DFT of length L_{FFT} involves $L_{\text{FFT}} \log_2 L_{\text{FFT}}$ real additions and real multiplications, which is valid with the radix-2 FFT algorithm if L_{FFT} is a power of two (otherwise, mixed radix FFT algorithms, whose complexity is also in $\mathcal{O}(L_{\text{FFT}} \log L_{\text{FFT}})$, may be employed [69]). First, the input vectors are transformed to the DFT domain, $2L_{\text{FFT}} \log_2 L_{\text{FFT}}(C_A + C_M)$ operations are involved. Then, the convolution is computed in the DFT domain with $\frac{L_{\text{FFT}}}{2} + 1$ complex multiplications, exploiting the Hermitian symmetry of the DFT for real-valued time-domain signals. Each complex multiplication requires four real multiplications and two real additions. Eventually, an IFFT is performed to obtain the result in the time domain. In total, the number of operations for the convolution is given by

$$C_{\text{conv}}^{L_{\text{FFT}}} = 3L_{\text{FFT}} \log_2 L_{\text{FFT}}(C_A + C_M) + (L_{\text{FFT}} + 2)(2C_M + C_A). \tag{8.22}$$

8.2.1 NLMS Complexity

Block-wise batch implementation

The NLMS algorithm in (4.7) may be implemented in a batch fashion with

$$\mathbf{a}^{(n+1)} = \mathbf{a}^{(n)} - \mu \frac{\widehat{\mathbf{r}}_{y\mathbf{x}_B}^{(n)}}{\|\mathbf{x}_B\|^2}, \tag{8.23}$$

where n represents the iterations index. The $M'L \times 1$ vector $\widehat{\mathbf{r}}_{y\mathbf{x}_B}^{(n)}$ is an estimator for the correlation $\mathbf{E}\left\{y^{(n)}\mathbf{x}_B(p)\right\}$. The computation of $y^{(n)}(p)$ on L points requires M' convolutions and $M'L$ additions. For simplicity and fairness, we may consider the same biased estimator as for our NG-SOS-BSS algorithms which was given in (6.47). Then the estimate $\hat{\mathbf{r}}_{yx_B}^{(n)}$ may be implemented with M' block convolutions. Taking the normalization by $\|\mathbf{x}_B\|^2$ into account, the number of operations in (8.23) is given by

$$C_{\mathrm{NLMS}}^{\mathrm{batch}}(M', L) = 2M' C_{\mathrm{conv}}^{2L} + (2M'L + 1)C_M + M'(2L - 1)C_A \quad (8.24)$$
$$= \mathcal{O}(L \log_2 L). \quad (8.25)$$

The complexity for a block-wise batch NLMS algorithm depends on the parameters N_{iter} and β. It is directly obtained by substituting C^{batch} in (8.21) with its value in (8.24). Thus, the complexity per sample is in $\mathcal{O}(\log_2 L)$.

Sample-wise implementation

The sample-wise implementation of the NLMS algorithm in (4.7) firstly requires the output sample $y(p)$ to be computed as $y(p) = x_0(p - D) + \mathbf{a}^{\mathrm{T}}(p)\mathbf{x}_B(p)$, which necessitates $M'L$ multiplications and $M'L$ additions. Once the output $y(p)$ is available, computing $y(p)\mathbf{x}_B(p)$ requires $M'L$ multiplications. We also need to take into account the normalization term, which may be computed efficiently as

$$\|\mathbf{x}_B(p)\|^2 = \|\mathbf{x}_B(p-1)\|^2 - \sum_{m=1}^{M'} |x_{B,m}(p - L)|^2 + \sum_{m=1}^{M'} |x_{B,m}(p)|^2. \ (8.26)$$

This requires $2M'$ multiplications and $2M'$ additions. Denote the complexity of a multiplication by C_M and that of an addition by C_A (divisions are treated as multiplications). Given the division by $\|\mathbf{x}_B(p)\|^2$ the subtraction in (4.7), the NLMS complexity becomes

$$C_{\mathrm{NLMS}}(M', L) = 2(M'(L + 1) + 1)C_A + 2M'(L + 1)C_A \quad (8.27)$$
$$= \mathcal{O}(L). \quad (8.28)$$

8.2.2 Complexity of NG-SOS-BSS Algorithms

This section derives the complexity of NG-SOS-BSS algorithms and distinguishes three cases: block-wise batch BSS, block-wise batch PBSS, and sample-wise PBSS in the case $N = 2$.

Block-wise batch implementation for a square system

Let us start with the derivation of the complexity for a square system with $M = N$. We rewrite the NG-SOS-BSS algorithm in its generic form:

$$\mathbf{W}(n + 1) = \mathbf{W}(n) - \mu \sum_{k=1}^{K} \mathrm{diag}^{-1}\left(\mathbf{R_{yy}}(kL)\right) \mathrm{boff}\left(\mathbf{R_{yy}}(kL)\right) \mathbf{W}(n). \quad (8.29)$$

Due to the Sylvester structure of $\mathbf{W}$, the computation of $\mathbf{y}(kL)$ in

$$\mathbf{y}(kL) = \mathbf{W}\mathbf{x}(kL), \qquad k = 1, \ldots, K, \tag{8.30}$$

actually consists of block convolutions. We may exploit the efficiency of the convolution in the DFT domain in particular if we transform the K input data blocks in one pass. The number of operations involved in (8.30), which is denoted by C_1, is then given by

$$C_1(N, L, K) = N^2 C_{\text{conv}}^{\max\{2L, KL\}} + KN(N-1)LC_A. \tag{8.31}$$

Once the output signals are available, the following operations are carried out:

(i) The cross-correlation matrices boff$(\mathbf{R}_{yy}(kL))$ for $k = 1, \ldots, K$ are estimated. We consider the biased estimation given by $r_{ij,\tau}(kL)$ in (6.47). We note that an unbiased estimation may also be considered and may have a larger complexity. Using blocks of output signal $\mathbf{y}_n(p)$ with length L, the cross-correlation estimator $r_{ij,\tau}(kL)$ for $\tau = -L+1, \ldots, L-1$ may be computed efficiently with convolutions in the DFT domain. In total, the estimation of boff$(\mathbf{R}_{yy}(kL))$ has complexity

$$C_2(N, L, K) = \frac{1}{2}KN(N-1)C_{\text{conv}}^{2L}. \tag{8.32}$$

(ii) The product boff$(\mathbf{R}_{yy}(kL))\,\mathbf{W}$ may also be implemented with block convolutions in the DFT domain. Strictly speaking, one should distinguish between (1) the self-closed rule, which may be computed with an FFT length of $2L$ and (2) the non-self-closed rule, which can be computed with an FFT length of $3L$. For the sake of simplicity, we consider only the self-closed rule. One finds that $KN^2(N-1)$ convolutions are required, with the complexity

$$C_3(N, L, K) = N^2(N-1)C_{\text{conv}}^{\max\{2L, KL\}} + KN^2(N-2)LC_A. \tag{8.33}$$

(iii) By averaging over L samples, the estimates $\|\mathbf{y}_n(kL)\|^2$ of the output signal powers for $k = 1, \ldots, K$ involve KNL multiplications and $KN(L-1)$ additions. Multiplying $\mathbf{w}_{nm}$ by $\mu/\|\mathbf{y}_n(kL)\|^2, n, m = 1, \ldots, N$ has complexity $KN(1 + NL)C_M$. Thus, normalizing by $\|\mathbf{y}_n(kL)\|^2$ for $n = 1, \ldots, N$ and for $k = 1, \ldots, K$ has complexity

$$C_4(N, L, K) = KN(L + NL + 1)C_M + KN(L-1)C_A. \tag{8.34}$$

Adding in the subtraction in (8.29) $(C_5(N, L, K) = KN^2LC_A)$, the total number of operations for one NG-SOS-BSS batch iteration is given by

$$C_{\text{BSS}}^{\text{batch}}(N, L, K) = \sum_{i=1}^{5} C_i(N, L, K). \tag{8.35}$$

Then the complexity in block-wise batch online implementation can be written using (8.21) as

$$C_{\text{BSS}}^{\text{block-wise}}(N, L, K) = \frac{N_{\text{iter}}}{\beta L} C_{\text{BSS}}^{\text{batch}}(N, L, K) \tag{8.36}$$

$$= \mathcal{O}(\log_2 L). \tag{8.37}$$

Block-wise batch implementation for a nonsquare system and $N = 2$

Consider the PBSS update given in (5.52) and assume that the cross filters $\mathbf{w}_{nm}, n \neq m, n, m > 1$ are not adapted as in (5.53). In this case the number of operations is given by

$$C_1^{\text{PBSS,batch}}(M, L, K) = 3(M - 2)KC_{\text{conv}}^{\max\{2L, KL\}} + 2(M - 1)KLC_A, \tag{8.38}$$

$$C_2^{\text{PBSS,batch}}(M, L, K) = K(M - 1)C_{\text{conv}}^{2L}, \tag{8.39}$$

$$C_3^{\text{PBSS,batch}}(M, L, K) = 4(M - 1)C_{\text{conv}}^{\max\{2L, KL\}} + (M - 2)KLC_A, \tag{8.40}$$

$$C_4^{\text{PBSS,batch}}(M, L, K) = KM(L + ML + 1)C_M + KM(L - 1)C_A, \tag{8.41}$$

$$C_5^{\text{PBSS,batch}}(M, L, K) = K(3M - 2)LC_A, \tag{8.42}$$

$$C_{\text{PBSS}}^{\text{batch}}(M, L, K) = \sum_{i=1}^{5} C_i^{\text{PBSS,batch}}(M, L, K). \tag{8.43}$$

(For each index $i = 1, \ldots, 5$, the variable $C_i^{\text{PBSS,batch}}(M, L, K)$ gives the complexity of the same algorithm part as the variable $C_i(N, L, K)$ above.) Then the number of operations per sample in block-wise batch online implementation can be written using (8.21) as

$$C_{\text{PBSS}}^{\text{block-wise}}(M, L, K) = \frac{N_{\text{iter}}}{\beta L} C_{\text{PBSS}}^{\text{batch}}(M, L, K), \tag{8.44}$$

$$= \mathcal{O}(\log_2 L). \tag{8.45}$$

It may be easily verified that $C_{\text{BSS}}^{\text{batch}}$ in (8.35) and $C_{\text{PBSS}}^{\text{batch}}$ in (8.43) are equal if $M = N = 2$:

$$C_{\text{BSS}}^{\text{batch}}(2, L, K) = C_{\text{PBSS}}^{\text{batch}}(2, L, K). \tag{8.46}$$

Sample-wise implementation for $N = 2$

In the case $N = 2$, sample-wise PBSS updates have been derived in (6.68). The computational cost results from

(*i*) the computation of M outputs $y_m(p)$ for $m = 1, \ldots, M$,

$$C_1(M, L) = M(M - 1)LC_M + M((M - 1)(L - 1) + 1)C_A, \tag{8.47}$$

(*ii*) the computation of $2(M - 1)$ cross correlations $y_m(p)\mathbf{y}_{m'}(p)$ for $m \neq m'$,

$$C_2(M, L) = 2(M - 1)LC_M, \tag{8.48}$$

(iii) the M normalization terms $\|\mathbf{y}_m(p)\|^2$ for $m = 1, \ldots, M$ as $\|\mathbf{y}_m(p)\|^2 = \|\mathbf{y}_m(p-1)\|^2 - y_m^2(p-L) + y_m^2(p)$,

$$C_3(M, L) = 2M(C_A + C_M), \tag{8.49}$$

(iv) $2(M-1)$ divisions by $\|\mathbf{y}_m(p)\|^2$, multiplication by μ, and eventual subtraction

$$C_4(M, L) = 2(M-1)LC_A + 2MC_M. \tag{8.50}$$

The total number of operations per sample is given by

$$C_{\text{PBSS}}(M, L) = \sum_{i=1}^{4} C_i(M, L) \tag{8.51}$$

$$= \mathcal{O}(L). \tag{8.52}$$

8.2.3 Comparison of NLMS and NG-SOS-BSS Complexities

Figure 8.3 compares the complexities for the NLMS and the NG-SOS-BSS in their sample-wise and block-wise implementations for the setup of $N = 2$ sources and $M = 4$ sensors. Note that this represents only the computational demands. (The performance of the algorithms will be measured from experiments and is taken into consideration in Sect. 8.4.) To obtain the minimum complexity for NG-SOS-BSS, the number of blocks is set to $K = 1$ and the number of iterations per batch run is set to $N_{\text{iter}} = 1$. The number of interference references for the NLMS is set to $M' = M - 1 = 3$, which corresponds to the number of adaptive filters in a standard GSC[4]. It may be observed that NG-SOS-BSS requires more operations per input sample than NLMS for both sample-wise and block-wise implementations. As expected, the use of the FFT for the block-wise implementation significantly reduces the complexity, in particular for long separation filters. For $L = 256$, the block-wise batch PBSS algorithm would require about 3,500 operations per sample, which is comparable to the complexity for a sample-wise NLMS-driven GSC (about 3,000 operations per sample). The price for this low complexity is the reduced tracking capability.

In the square case ($N = M = 2$), the number of operations for the sample-wise NG-SOS-BSS update in (6.67) can be obtained by setting $M = 2$ in (8.51), which results in

$$C_{\text{BSS}}(2, L) = 4(L+2)C_M + 4(L+1)C_A. \tag{8.53}$$

---

[4] The case of the RGSC would be obtained by setting $M' = M = 4$, which also accounts for the filters in the adaptive blocking matrix since adaptation of the blocking matrix and of the interference canceler never occur simultaneously (see Appendix C for more details). This would result in approximately $\frac{M}{M'} = \frac{4}{3}$ more operations than for the GSC.

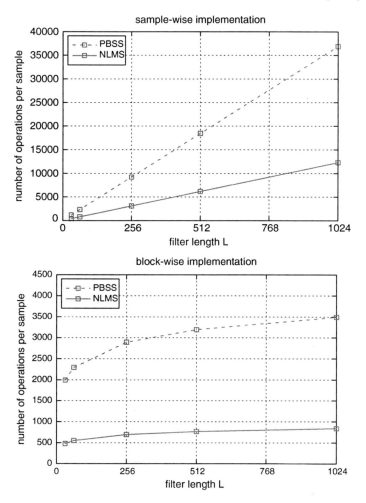

Fig. 8.3. Number of operations (sum of the number of real additions and of the number of real multiplications) for $M = 4, M' = M - 1, N = 2, K = N_{\text{iter}} = 1$ for (*Upper plot*): sample-wise implementation and (*lower plot*): block-wise implementation

This is exactly twice as many operations per sample compared to the NLMS algorithm in (4.7) for $M' = 1$, which is given by

$$C_{\text{NLMS}}(1, L) = 2(L + 2)C_M + 2(L + 1)C_A. \qquad (8.54)$$

In other words, the sample-wise NG-SOS-BSS update in (6.67) and the sample-wise NLMS algorithm in (4.7) require the same number of operations *per output sample* (which account for the fact that the sample-wise NG-SOS-BSSupdate in (6.67) achieves "more than" the NLMS algorithm since it provides estimates not only for the target signal but also for the interference signal).

8.3 Links with the ILMS Algorithm

As we have seen above, the sample-wise NG-SOS-BSS update in (6.68) and the sample-wise NLMS algorithm in (4.7) are very similar in terms of computational load for $M = 2$. In fact, tight connections between (6.68) and the ILMS algorithm exist.

Let us rewrite the ILMS algorithm in (4.8) in terms of $\mathbf{w}_{12}$ for the AIC architecture in Fig. 5.1:

$$\mathbf{w}_{12}(p+1) = \mathbf{w}_{12}(p) - \mu \frac{y_1(p)\mathbf{x}_2(p)}{\|\mathbf{y}_1(p)\|^2}. \tag{8.55}$$

Following the suggestion of Van Gerven [38], we replace the interferer reference signal $\mathbf{x}_2(p)$ by its best estimation $\mathbf{y}_2(p)$ available in BSS. Then (8.55) becomes

$$\mathbf{w}_{12}(p+1) = \mathbf{w}_{12}(p) - \mu \frac{y_1(p)\mathbf{y}_2(p)}{\|\mathbf{y}_1(p)\|^2}, \tag{8.56}$$

which is just the NG-SOS-BSS algorithm in (6.67) for $(n, m) = (1, 2)$ in the case $\|\mathbf{y}_1(p)\|^2 > 2\alpha$. This link between NG-SOS-BSS and ILMS suggests that NG-SOS-BSS exploits the source signal nonstationarity in a particular way by taking special advantage of the source silences (as ILMS does). For both algorithms, the parameter updates have the largest *norm* during silences of the target source. Note, however, that the *directions* of their updates are different.

Another line of comment about the ILMS algorithm is in order. We observe that replacing the normalization term $\tilde{r}_{y_n y_n, 0}(p)$ with 1, (6.67) reduces to

$$\mathbf{w}_{12}(p+1) = \mathbf{w}_{12}(p) - \mu y_1(p)\mathbf{y}_2(p), \tag{8.57}$$

$$\mathbf{w}_{21}(p+1) = \mathbf{w}_{21}(p) - \mu y_2(p)\mathbf{y}_1(p), \tag{8.58}$$

which is simply the algorithm (5.3) proposed by Van Gerven et al. that we mentioned in the introduction of Chap. 5. Van Gerven's algorithm appears as an approximation of the sample-wise NG-SOS-BSS algorithm in the case $N = 2$. From the analogy between ILMS and NG-SOS-BSS, we can see that Van Gerven was not far away from an efficient BSS algorithm. He only missed an adequate normalization. In fact, he also derived a normalized update in [39]. Starting from the normalization by $\|\mathbf{x}_2(p)\|^2$ that is applied in the NLMS algorithm, Van Gerven proposed to normalize (8.57) by the update $\|\mathbf{y}_2(p)\|^2$, as follows

$$\mathbf{w}_{12}(p+1) = \mathbf{w}_{12}(p) - \mu \frac{y_1(p)\mathbf{y}_2(p)}{\|\mathbf{y}_2(p)\|^2}. \tag{8.59}$$

Unfortunately, this normalization is counterproductive. In (8.59), the adaptation becomes slow when the interferer is dominant.

8.4 Experimental Comparison

In this section, we want to compare NG-SOS-BSS and LMS-adapted LCMV beamforming in terms of performance-to-complexity ratio. From a practical standpoint, online algorithms will be analyzed, as opposed to offline algorithms.

Experimental conditions

Recordings of real speakers in a car cabin are used, as in Sect. 6.3. We refer to Appendix A for more details on the experimental setups.

Algorithms and parameter settings

The following algorithms are considered:

- *LMS-adapted LCMV beamforming algorithms.* We use the implicitly controlled **ILMS** and the **DTD-NLMS** implemented as described in Sect. 4.5. For the **DTD-NLMS**, adaptation is carried out during periods of interference activity and target silences, which requires a double-talk detector (DTD). The adaptation control stops the adaptation when the power of the true target signal exceeds a fixed threshold. (In practice, the true target signal in unknown. Hence, this adaptation control may be regarded as a best-case double-talk detection.) Tables 4.1 and 4.2 may be referred for more details on the implementation.
- *NG-SOS-BSS algorithms.* We consider the **sample-wise** adaptive algorithm (6.67). **Block-wise** adaptation is performed with the self-closed update (6.51) if $M = 2$ and with PBSS algorithm (6.54) if $M = 4$. In all cases, a causal separation system is used ($d = 0$). As described in Sect. 6.2.1, the computation load of block-wise online NG-SOS-BSS algorithm depends on the number of blocks K, the number of iterations N_{iter}, and the frame rate β. To limit the number of necessary experiments, we assume that these parameters influence the performance independently of each other and we let them vary separately from the point $(K, N_{\text{iter}}, \beta) = (1, 1, 1)$ and consider the following parameter settings

$$
\begin{array}{c|c|c|c}
K = & 1, \ldots, 5 & 1 & 1 \\
N_{\text{iter}} = & 1 & 1, \ldots, 5 & 1 \\
\beta = & 1 & 1 & \frac{1}{5}, \frac{1}{4} \ldots, \frac{1}{1}
\end{array}
\qquad (8.60)
$$

The filter length is also set to $L = 256$. Tables 6.3 and 6.4 may be referred for more details on the implementation.

It may be difficult to compare different algorithms fairly. In particular, the step-size parameter influences the performance significantly. We arbitrarily decided not to change the step-size parameters and set them as in Sect. 4.5 for LCMV beamforming and as in Sect. 6.3 for NG-SOS-BSS algorithms. In practice, the performance of adaptive LCMV beamforming is also particularly dependent on the implementation of the control mechanism.

Performance and complexity measures

To compare SOS-BSS and LMS-adapted LCMV beamforming in terms of performance and complexity, we consider the following objective measures:

- The performance measures are the start-up SIR improvement $Q_{[0,3]}$ and the SIR improvement after initial convergence $Q_{[3,10]}$, as defined in Table 2.1.
- The complexity measure is the number of FLoating point OPerations (real additions and multiplications) per Second in Millions (MFLOPS) evaluated using the results from Sect. 8.2. We did not account for the fact that the adaptation of the DTD-NLMS algorithm is carried out only during periods of time with interference activity and target silence, which reduces the number of necessary operations compared to unsupervised methods (but the difference might be counterbalanced by the computational load for the DTD).

The results are shown in Figs. 8.4 and 8.5.

First, let us compare NG-SOS-BSS algorithms to the supervised beamforming algorithm DTD-NLMS. As may be seen in Fig. 8.4a and Fig. 8.5a, the superior start-up performance of DTD-NLMS clearly appears for both microphone setups. This may be explained by the conjunction of two factors. First, the source signals have little overlap in the first half of the recordings. Second, NLMS may converge very fast if the step-size is large. By contrast, a large step-size makes NG-SOS-BSS algorithms unstable.

Then, in the second part of the recordings, the performance of DTD-NLMS depends on the microphone setup. With the four-element compact array mounted in the rear-view mirror, the intermittent adaptation is sufficient to reduce the interferer also during double-talk (Fig. 8.4b). In this case, the performance of DTD-NLMS is similar to that of PBSS. On the other hand, with the two-element distributed array mounted on the car ceiling, the performance is rather limited during double-talk (Fig. 8.5b). Here, continuous adaptation seems necessary and the performance of NG-SOS-BSS is superior, even with a similar computational load as may be seen from Fig. 8.5b for $K = N_{\text{iter}} = \beta = 1$. This discrepancy may be due to the fact that the interference signal suppression relies more on spatial than on spectral filtering in the case of the four-element compact array mounted in the rear-view mirror. By contrast, for the two-element distributed array mounted on the car ceiling, the interference signal suppression may rely more on spectral than on spatial filtering.

Now, we compare NG-SOS-BSS to the implicitly controlled beamforming algorithm ILMS. Here again, the superior start-up performance of ILMS appears for both microphone setups. As we may see in Fig. 8.5a,b, the sample-wise NG-SOS-BSS algorithm and ILMS yield similar performances with the two-element distributed array mounted on the car ceiling. This may be expected from their relationship described in Sect. 8.3. However, for the four-element compact array mounted in the rear-view mirror, the score $Q_{[3,10]}$ of

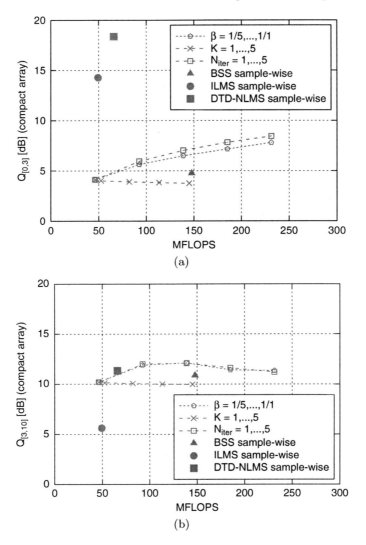

Fig. 8.4. Performance as a function of the computational load for LMS-adapted LCMV beamforming and NG-SOS-BSS algorithms for the four-element compact array mounted in the rear-view mirror

ILMS is poor (Fig. 8.4b). This reflects the target signal cancelation that we reported in Sect. 4.5.1. By contrast, PBSS does not reduce the target signal.

The following observations on NG-SOS-BSS may be done:

- In all cases, increasing the number of blocks K improves the performance at best marginally.
- With the two-element distributed array mounted on the car ceiling, sample-wise NG-SOS-BSS outperforms the block-wise adapted algorithms,

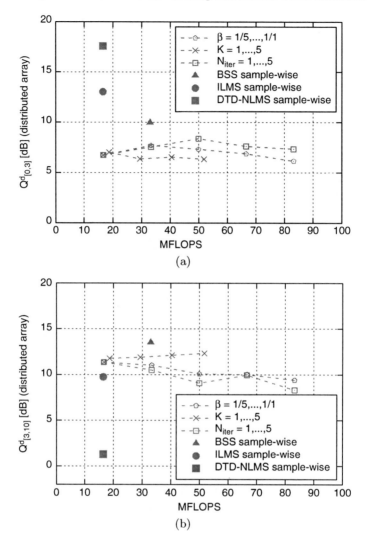

Fig. 8.5. Performance as a function of the computational load for LMS-adapted LCMV beamforming and NG-SOS-BSS algorithms for the two-element distributed array mounted on the car ceiling

in terms of performance-to-complexity ratio. This is because the optimum separation filters depend on the signals' spectral content. Hence, the performance depends on the tracking capabilities of the adaptation. By contrast, with the four-element compact array mounted in the rearview mirror, the performances of sample-wise and block-wise algorithms are similar.

Table 8.2 summarizes these experimental results.

Table 8.2. Results of the experimental comparison of SOS-BSS and adaptive beam-forming

	LMS-adapted LCMV beam-forming	NG-SOS-BSS
adaptation	controlled by a DTD (except for ILMS)	unsupervised
start-up performance	depends on the DTD, on the amount of double-talk and on the step-size. Fast convergence capability	slow because the global stability is not guaranteed, which requires a small step-size
performance after initial convergence	depends on the DTD, on the amount of double-talk	depends on the validity of the LTI mixing model

8.5 Summary and Conclusion

In the first part of this chapter, we discussed theoretical properties of SOS-BSS and LCMV beamforming. First, the global shapes of the cost functions have been examined. For LCMV beamforming, the cost function J_{LS} has a unique minimum, and the convergence of the gradient descent to this minimum can be guaranteed if the step-size is within a range known a priori. The situation is much more complex in the case of SOS-BSS. The cost function J_{BSS} has multiple minima. Setting the causality constraint on the separation matrix limits the search space a priori and removes the minima that correspond to permuted solutions. The convergence of the natural gradient for J_{BSS} cannot be guaranteed yet, even in the case of instantaneous mixtures. The conditions for interference-independent separating solutions to be minimum of J_{BSS} are less restrictive than for J_{LS}.

The cost functions have then been discussed from an estimation point of view. In a simple case with instantaneous mixtures, we could show that both techniques are superefficient if the source of interest has periods of silence. By contrasting with another widely used BSS technique based on the second-order statistics, we could explain why the mutual-information-based cost function may be superefficient.

The second part of this chapter compared the two approaches from a practical standpoint. First we derived the computational complexities for LMS-adapted LCMV beamforming and NG-SOS-BSS for block-wise and sample-wise updates. We have shown that the sample-by-sample update proposed in Sect. 6.2.2 is similar to the NLMS algorithm in terms of complexity. We also showed that this sample-by-sample update is very closely linked to the ILMS algorithm given in Chap. 4.

Finally, NG-SOS-BSS algorithms and LMS-adapted LCMV beamforming have been compared in terms of performance-to-complexity ratio. On the one hand, it appears that the NG-SOS-BSS algorithms have a slower initial convergence than LCMV beamforming. On the other hand, they may outperform

LMS-adapted LCMV beamforming during double-talk periods. This is true even in the particular case of the two-element distributed array mounted on the car ceiling, where LCMV beamforming adaptation may be performed continuously with the ILMS algorithm. The experiments have confirmed the good performance of the NG-SOS-BSS sample-by-sample update when rapid tracking of the signal spectrum is required. This shows that NG-SOS-BSS algorithms are a true alternative to LMS-adapted LCMV beamforming.

To summarize, the most important results of this chapter are:

- The convergence of the gradient descent for the LCMV cost function can be guaranteed, while the convergence of the SOS-BSS natural gradient cannot be guaranteed yet.
- Source silences appear to play an important role in adaptive LCMV beamforming *and* in SOS-BSS, as shown by the link between the ILMS algorithm and the sample-wise NG-SOS-BSS algorithm. In both cases, source silences enable statistical superefficiency.
- NG-SOS-BSS algorithms converge slower but may outperform LMS-adapted LCMV beamforming because the former are able to adapt during double-talk.

9
Combining Second-Order Statistics BSS and LCMV Beamforming

Adaptive LCMV beamforming and convolutive blind source separation (BSS) have a common goal, namely to reduce interferences. On the one hand, BSS algorithms are able to adapt continuously, while LCMV beamforming algorithms adapt only when the interferer signal is dominant. On the other hand, adaptive LCMV beamforming algorithms may converge faster than NG-SOS-BSS algorithms if there is no double-talk, as we have observed in Chap. 8. Moreover, beamformers may exploit geometric prior information about the position of the target source. In this chapter our objective is to combine the advantages of both approaches.

A combination of a BSS algorithm and a beamforming algorithm depends on several design options. At first glance, the beamformer could be either data-independent or adaptive, or it could be placed either in front of or after the BSS block, leading to four possible combinations. However, two combinations may be removed a priori. Firstly, placing a data-independent beamformer after the BSS block is not appropriate. This is because the BSS separation system distorts the spatial properties of the signals on which data-independent beamformers are based. Secondly, the combination of an adaptive beamformer in front of the BSS block can also be removed a priori. The adaptive LCMV beamformer is generally highly time varying. The BSS separation system could bring an improvement if it was able to track the time variance of the beamformer, that is, if it converged faster than the beamformer. This is generally not the case, as shown in Chap. 8.

As a result, two combinations deserve further investigation: (1) the data-independent beamformer in front of a BSS adaptive block, and (2) the adaptive beamformer as a postprocessing of the BSS output signals. Essentially, combination (1) is a particular way of integrating geometric prior information in BSS and will be examined in Sect. 9.2. Combination (2) will be studied in Sects. 9.3 and 9.4. Finally, Sect. 9.5 will compare the obtained algorithms as a front-end in automatic speech recognition.

9.1 Existing Combinations

Considering the two aspects of beamforming (geometric prior information about the position of the target source, and the LMS power criterion for the adaptation of the interference canceler), we may classify the existing approaches into two classes:

- Different approaches have been proposed to improve BSS with geometric information on the source directions of arrival (DOA). Based on the interpretation of BSS as a set of null beamformers, a one-to-one correspondence between the output signals and the DOA of the sources can be determined [10]. This allows to resolve the permutation ambiguities. In the case of frequency-domain BSS algorithms, where the permutation problem is critical, this may significantly improve the performance [58, 63, 82]. The influence of DOA information may be made stronger than for solving the permutations only. Saruwatari et al. have proposed to replace the BSS separation system with null beamformers if their outputs are less correlated [80]. This approach is well-suited if the information on the source position is reliable and if the far- and free-field propagation model matches with the actual acoustic environment. By contrast, in reverberant environments, BSS may outperform a data-independent null beamformer, even if the position of the sources is precisely known [58]. Initialization of the separation system to null beamformers was also proposed for time-domain algorithms [3]. A general approach for merging frequency-domain BSS and geometric prior information was presented by Parra et al. [77]. They proposed to use the geometric information at the initialization of the separation system or as a soft constraint. Their experimental evaluation showed that the simple initialization may lead to the best performance. Another use of prior geometric information in the field of BSS appears with sparsity-based techniques: One may assume that, even though the sources overlap in the time domain, they do not overlap in an auxiliary transform domain. In practice, the STFT is chosen as an auxiliary transform, and the sparsity assumption says that the sources are not active at the same points in the time-frequency plane so that they may be separated spectrally. Geometric information may be used to determine which points in the time-frequency plane belong to which source. This approach may lead to very efficient algorithms [1, 15, 78]. However, the sparsity assumption will usually not be entirely satisfied and sparsity-based techniques may distort the source signals.
- The second class of combinations of source separation and beamforming consist of using the LMS power criterion for adapting the filters of the separation system. For example, the use of the LMS power criterion was proposed as a postprocessing to cancel the residual crosstalk components in the BSS-separated signals [67, 72, 87].

In this chapter, we build upon the time-domain self-closed NG-SOS-BSS algorithms (6.51) and (6.54) and investigate how geometric prior information or the LMS power criterion may be efficiently combined with BSS. The different options are discussed and lead to both known and novel broadband algorithms.

9.2 BSS and Geometric Prior Information

This section discusses how geometric prior information may be used in BSS. First, Sect. 9.2.1 investigates the impact of causality constraints. These may be set if the sources are on different sides of the microphone median plane (see Fig. 6.2) but they do not require a precise information on the source DOA. Second, in Sects. 9.2.2–9.2.5, we present and discuss techniques that allow to apply geometric constraints on the separation system. The performance improvement that is due to the input of geometric prior information depends on the accuracy of this prior information in the current experimental conditions. In this section, the geometric prior information consists of the source DOA and the geometric constraints are based on the far- and free-field acoustic propagation model, which is only an approximation of the physical reality. It may not be desirable to apply these constraints strictly, since the loss of the number of degrees of freedom might not be compensated by the contribution of the geometric information [58]. This section presents techniques to enforce a given geometric constraint with various degrees of strictness:

- The constraint may be set only at the initialization. This is discussed in Sect. 9.2.3.
- Intermediately, the geometric prior information may be included as a soft constraint, which is maintained with a parameterizable degree of strictness (Sect. 9.2.4).
- The constraint may be strictly maintained during the entire adaptation (Sect. 9.2.5).

More realistic models may be considered, possibly requiring more prior information than the source DOA. Conversely, simulations could be carried out in a controlled acoustical environment with varying reverberation. Here the experiments are conducted in the car interior which is our privileged application environment. Also, the geometric prior information may be included *at the physical level*, for example by directing the microphone toward the sources as it is done with the distributed microphone array.

Experimental evaluation

In this section we are interested in merging NG-SOS-BSS algorithms with prior information at the algorithmic level in the car environment. The experimental setup is described in Appendix A with the four-element compact array mounted in the rear-view mirror. The geometric prior information is set to $(\theta_1, \theta_2) = (20°, -40°)$, which may be regarded as a fairly accurate estimation of the true DOA.[1]

The performance is measured with $Q_{[0,3]}$ and $Q_{[3,10]}$ as defined[2] in Table 2.1. All experimental results in this chapter are obtained for $L = 256$. As a baseline, the NG-SOS-BSS algorithm (6.54) is used with a causal separation system ($d = 0$). The parameters settings are summarized in Table 6.4.

9.2.1 Causality Information

There is a rather weak geometric prior information that nevertheless plays a significant role in BSS. It is the information whether or not a causal separation system may be used, as depicted in Fig. 6.2. For the separation of the driver and the codriver in the car cabin, the microphone arrangement mounted in the rear-view mirror is so that the source signals may be separated by a causal separation system. That is, the sources are on different sides of the microphone median plane. For this reason, we initialize the separation system with $\overline{\mathbf{W}}(z, 0) = \mathbf{I}$ as in (6.69) and use the self-closed update (6.54) with $d = 0$.

In the following, we investigate the performance of BSS algorithms when this information is *not* available by setting the initial separation system to $\overline{\mathbf{W}}(z, 0) = \mathbf{I}z^{-L/2}$. We then adapt $\mathbf{W}$ with the self-closed update in (6.54) and $d = L/2$ (the other parameters are given in Table 6.4). The performance is reported in Table 9.1. For the sake of comparison, the performance of the frequency-domain BSS algorithms from Sect. 6.3.3 is also given. For these algorithms, the learning rules remain unchanged, the distinction between causal and acausal separation system is made only at the initialization in (6.69) for $d = 0$ and $d = L/2$, respectively.

As we can see in Table 9.1, the performance of all algorithms decreases when the causality information is not used. However, the time-domain baseline algorithm (6.54) exhibits a relatively robust behavior: Its performance

Table 9.1. Comparison of the NG-SOS-BSS baseline and frequency-domain BSS algorithms with the four-element compact array mounted in the rear-view mirror (see Table 6.4 and Sect. 6.3.3 for more details on the algorithm implementation)

	NG-SOS-BSS		FD-SOS		FD-HOS	
performance	$Q_{[0,3]}$ (dB)	$Q_{[3,10]}$ (dB)	$Q_{[0,3]}$ (dB)	$Q_{[3,10]}$ (dB)	$Q_{[0,3]}$ (dB)	$Q_{[3,10]}$ (dB)
causal (baseline)	4.0	10.2	4.1	6.9	4.1	7.8
acausal	2.5	9.1	1.5	4.4	0.8	2.1

---

[1] There is also an influence from array imperfections. Since these are not controlled in practice, we regard their influence as a part of the influence of the reverberation.

[2] $Q_{[t_1, t_2]}$ measures the SIR improvement averaged over the time interval $[t_1, t_2]$, t_1 and t_2 in seconds.

after initial convergence, $Q_{[3,10]}$, decreases by only 1.1 dB when an acausal separation system is used, representing a 10% relative performance loss. The performance of the frequency-domain BSS algorithms decreases more severely, for example, $Q_{[3,10]}$ decreases by more than 2 dB, representing at least 30% relative performance loss.

The breakdown of the frequency-domain BSS algorithms may be explained by the permutation problem: In case of successful but permuted separation, a full, broadband permutation of the output channels would yield in large SR and small IR values, i.e., negative values for $Q_{[0,3]}$ and $Q_{[3,10]}$. If permutations occur inconsistently across the frequency range, the SIR improvement may be large at frequencies where no permutation occurred and small elsewhere. In average, this results in smaller values for $Q_{[0,3]}$ and $Q_{[3,10]}$, depending on the number of inconsistent permutations.

By contrast, the permutations are inherently prevented in the causal scenario. More recently, the role of causality in frequency-domain BSS was observed in other experiments by Robledo-Arnuncio et al. [79] (see also [5]).

Again, it should be mentioned that the causality of the separation system is a matter of initialization. Setting the initial separation system to $\overline{\mathbf{W}}(z,0) = \mathbf{I}$ allows a causal mixing system to be separated, as depicted in Fig. 6.2a, but fails to separate an acausal mixing as in Fig. 6.2b. The proposal has been made to constrain the diagonal separation filters to unit responses, i.e. $\mathrm{diag}\overline{\mathbf{W}}(z,n) = \mathbf{I}$ for all n, as in the approach proposed by Parra et al. [75]. Parra introduces this constraint to neutralize the scaling ambiguity. However, this constraint not only neutralizes the scaling ambiguity, but also maintains the causality of the separation system. Parra reported good separation performance with an experimental setup as depicted in Fig. 6.2a, that is, the sources are on different sides of the microphone median plane. His approach has been applied later in different acoustic conditions by Ikram et al. [57]. Ikram found that Parra's algorithm did not perform satisfactorily, and explained this observation by the length of the room impulse response for his experimental conditions. In fact the experimental setup used by Ikram was on the borderline of requiring acausal separation filters, since one of the sources was placed *on* the microphone median plane. Therefore, the degree of difficulty for his experimental setup was higher than in [75]. Considering the causality of the mixing/separation system as in Fig. 6.2 gives further insights into the behavior and the performance of BSS algorithms.

9.2.2 Prior Information on the Source Direction of Arrival

This section formulates the geometric prior information in such a way that it may be included into time-domain BSS algorithms. We will reuse the space response, which has been introduced in Sect. 2.2.3. In the remainder of this chapter, the time index p in $\mathbf{W}(p)$ may be omitted for the sake of readability.

Spatial response

First, it is important to identify the separation matrix

$$\mathbf{W} = \begin{bmatrix} \mathbf{W}_{11} & \cdots & \mathbf{W}_{1M} \\ \vdots & \ddots & \vdots \\ \mathbf{W}_{N1} & \cdots & \mathbf{W}_{NM} \end{bmatrix} \tag{9.1}$$

as a set of N MISO systems. We denote each MISO system by $\mathbf{W}_n$, for $n = 1, \ldots, N$. That is, $\mathbf{W}_n$ is the nth row of $\mathbf{W}$:

$$\mathbf{W}_n \triangleq [\mathbf{W}_{n1}, \ldots, \mathbf{W}_{nM}]. \tag{9.2}$$

Then the output $\mathbf{y}_n(p)$ may be written as $\mathbf{y}_n(p) = \mathbf{W}_n\mathbf{x}(p)$. The wideband spatial response for $\mathbf{W}_n$ is denoted by $\mathbf{g}(\mathbf{W}_n, \theta)$ and is defined[3] as:

$$\mathbf{g}(\mathbf{W}_n, \theta) \triangleq \mathbf{W}_n\mathbf{d}(\theta). \tag{9.3}$$

The $M(2L-1) \times 1$ vector $\mathbf{d}(\theta)$ is given for an equispaced array by

$$\mathbf{d}(\theta) \triangleq \left[\mathbf{d}_1^{\mathrm{T}}(\theta), \ldots, \mathbf{d}_M^{\mathrm{T}}(\theta)\right]^{\mathrm{T}} \tag{9.4}$$

with
$$\begin{aligned} \mathbf{d}_m(\theta) &\triangleq (d_{m,L-1}(\theta), \ldots, d_{m,0}(\theta), \ldots, d_{m,-L+1}(\theta))^{\mathrm{T}} \\ d_{m,l}(\theta) &\triangleq \begin{cases} 1 & \text{if } l = (m-1)\tau_\theta, \quad \text{for } m = 1, \ldots, M. \\ 0 & \text{otherwise}, \end{cases} \end{aligned} \tag{9.5}$$

The delay τ_θ is the time needed by an acoustic wave with DOA θ to travel from one sensor to the next one. It is given by $\tau_\theta = f_s \Delta \sin(\theta)/c$, where Δ denotes the interelement spacing. In the definition (9.5), it is assumed that the delay τ_θ is an integer. If this is not the case, $\mathbf{d}(\theta)$ may be computed using fractional delay filters [64]. The vector $\mathbf{d}_m(\theta)$ should be seen as the microphone snapshot at time $p = L-1$, $\mathbf{x}_m(L-1)$, when a source in DOA θ emits a unit impulse at time $p = 0$. Then the $L \times 1$ vector $\mathbf{g}(\theta)$ is simply the output at time $p = L-1$, $\mathbf{y}(L-1)$.

Constraining the spatial response

We may control the spatial response of $\mathbf{W}_n$ at a particular DOA θ by constraining $\mathbf{g}(\mathbf{W}_n, \theta)$. For example, we may consider a constraint of unit response on $\mathbf{W}_n$ at a given DOA θ_0 with

$$\mathbf{g}(\mathbf{W}_n, \theta_0) = \boldsymbol{\delta}_d. \tag{9.6}$$

The integer d should be selected in the range $d = 1, \ldots, L$ and it controls the acausal length of the separation filters. Note that setting $d = L$ leads to causal

---

[3] This is simply a reformulation of the spatial response $\mathbf{g}(\mathbf{w}, \boldsymbol{\theta})$ given in (2.27) where the convolution is rewritten using a matrix product.

separation filters (no acausal coefficients). This is because the signal vectors $\mathbf{x}(p)$ and $\mathbf{y}(p)$ are time-reversed. The value $L - d$ represents the input–output delay. Setting $d = L/2$ leads to separation filters that have causal and acausal parts of equal lengths. The constraint in (9.6) is fulfilled in particular for delay-and-sum beamformers oriented at θ_0. In the following, delay-and-sum beamformers are used to implement constraints of unit spatial response.

If the DOA $\theta_1, \ldots, \theta_N$ of the sources $s_1, \ldots, s_N$ are known, we can express multiple geometric constraints

$$\mathbf{g}(\mathbf{W}_n, \theta_{n'}) = \mathbf{g}_{nn'}, \tag{9.7}$$

for $n, n' = 1, \ldots, N$. The $L \times 1$ vectors $\mathbf{g}_{nn'}$ determines how the spatial response of $\mathbf{W}_n$ is constrained at the DOA $\theta_{n'}$; they might be set arbitrarily. We may typically set $\mathbf{g}_{nn'} = \mathbf{0}$ for $n \neq n'$ and $\mathbf{g}_{nn} = \boldsymbol{\delta}_d$. Several geometric constraints can be stacked together for the whole separation system as follows:

$$\mathbf{WD} = \mathbf{G} \tag{9.8}$$
$$\text{with } \mathbf{D} \triangleq \begin{bmatrix} \mathbf{d}(\theta_1) \ldots \mathbf{d}(\theta_N) \end{bmatrix}, \tag{9.9}$$
$$\text{and } \mathbf{G} \triangleq \begin{bmatrix} \mathbf{g}_{11} & \cdots & \mathbf{g}_{1N} \\ \vdots & \ddots & \vdots \\ \mathbf{g}_{N1} & \cdots & \mathbf{g}_{NN} \end{bmatrix}. \tag{9.10}$$

The $NL \times N$ matrix $\mathbf{G}$ determines the geometric constraints. Note that the number of constraints is not necessarily equal to the number of sources. The response $\mathbf{g}(\mathbf{W}_n, \theta)$ of $\mathbf{W}_n$ for a particular DOA θ may be left unconstrained. In this case, we use the following notational device:

$$\mathbf{g}(\mathbf{W}_n, \theta) = \diamond. \tag{9.11}$$

This may be applied to the case of two sources and four microphones with partial blind source separation (PBSS). Since the outputs $\mathbf{y}_2, \mathbf{y}_3$ and $\mathbf{y}_4$ are assigned to the same source s_2, we can write (9.8) as:

$$\mathbf{WD} = \begin{bmatrix} \mathbf{g}_{11} & \mathbf{g}_{12} \\ \mathbf{g}_{21} & \mathbf{g}_{22} \\ \mathbf{g}_{21} & \mathbf{g}_{22} \\ \mathbf{g}_{21} & \mathbf{g}_{22} \end{bmatrix} \text{ with } \mathbf{D} = \begin{bmatrix} \mathbf{d}(\theta_1) \, \mathbf{d}(\theta_2) \end{bmatrix}, \tag{9.12}$$

where $\mathbf{g}_{nn} \in \{\boldsymbol{\delta}_d, \diamond\}$ for $n = 1, 2$ and $\mathbf{g}_{nm} \in \{\mathbf{0}, \diamond\}$ for $m \neq n$. By varying $\mathbf{g}_{nm}$ for $n, m = 1, 2$ in (9.12), 16 different constraints are generated. For later reference, we define two particular constraints $\mathbf{G}_0$ and $\mathbf{G}_1$ as:

$$\mathbf{G}_0 \triangleq \begin{bmatrix} \diamond & \mathbf{0} \\ \mathbf{0} & \diamond \\ \mathbf{0} & \diamond \\ \mathbf{0} & \diamond \end{bmatrix}, \quad \mathbf{G}_1 \triangleq \begin{bmatrix} \boldsymbol{\delta}_{L/2} & \diamond \\ \diamond & \diamond \\ \diamond & \diamond \\ \diamond & \diamond \end{bmatrix}. \tag{9.13}$$

The constraint $\mathbf{G}_0$ restricts $\mathbf{W}_n, n = 1, \ldots, 4$ to have a zero response in directions $\theta_{n'}$ for $n' \neq n$. On the other hand, $\mathbf{G}_1$ constrains $\mathbf{W}_1$ to have a unit response in the direction of the target source θ_1. We associate $\mathbf{G}_1$ with the delay-and-sum beamformer and the acausal length of the separation filters is set to $d = L/2$.

In the following, we present techniques to enforce a given geometric constraint with various degrees of strictness:

- In Sect. 9.2.3, the constraint is set only at the initialization, that is, on $\mathbf{W}(0)$.
- In Sect. 9.2.4, the geometric prior information is included as a soft constraint maintained with a parameterizable degree of strictness.
- In Sect. 9.2.5 the constraint is maintained strictly during the entire adaptation.

9.2.3 Geometric Information at the Initialization

Constraint $\mathbf{G}_0$

As shown by Araki et al., BSS-adapted separation systems converge to a set of null beamformers [10]. Thus, a straightforward way to use geometric prior information is to initialize the separation system with null beamformers [3]. In terms of geometric constraint in (9.8), this corresponds to setting $\mathbf{W}(p)$ at $p = 0$ so that

$$\mathbf{W}(0)\mathbf{D} = \mathbf{G}_0. \tag{9.14}$$

In the case $M = 4$, the constraint (9.14) is satisfied if the components of $\mathbf{W}(0)$ are set to

$$\begin{bmatrix} \mathbf{w}_{11}(0) & \mathbf{w}_{12}(0) & \mathbf{w}_{13}(0) & \mathbf{w}_{14}(0) \\ \mathbf{w}_{21}(0) & \mathbf{w}_{22}(0) & \mathbf{w}_{23}(0) & \mathbf{w}_{24}(0) \\ \mathbf{w}_{31}(0) & \mathbf{w}_{32}(0) & \mathbf{w}_{33}(0) & \mathbf{w}_{34}(0) \\ \mathbf{w}_{41}(0) & \mathbf{w}_{42}(0) & \mathbf{w}_{43}(0) & \mathbf{w}_{44}(0) \end{bmatrix} = \begin{bmatrix} \delta_0 & -\frac{1}{3}\delta_{-\tau_{\theta_2}} & -\frac{1}{3}\delta_{-2\tau_{\theta_2}} & -\frac{1}{3}\delta_{-3\tau_{\theta_2}} \\ -\delta_{\tau_{\theta_1}} & \delta_0 & \mathbf{0} & \mathbf{0} \\ -\delta_{2\tau_{\theta_1}} & \mathbf{0} & \delta_0 & \mathbf{0} \\ -\delta_{3\tau_{\theta_1}} & \mathbf{0} & \mathbf{0} & \delta_0 \end{bmatrix}. \tag{9.15}$$

This initialization preserves the causality of the separation system, which is possible for our experimental setup because $\tau_{\theta_1} > 0$ and $\tau_{\theta_2} < 0$. As may be seen from the results in Table 9.2, the initialization in (9.15) improves the performances $Q_{[0,3]}$ and $Q_{[3,10]}$.

Constraint $\mathbf{G}_1$

We may hope to improve the performance if, in addition to canceling the interfering signals, we also enhance the target signal. To this end, let us first initialize $\mathbf{W}_1$ as a delay-and-sum beamformer directed toward the target. This initialization requires an acausal separation system and may be realized as follows:

$$\mathbf{w}_{1m}(0) = \frac{1}{M}\boldsymbol{\delta}_{L/2-(m-1)\tau_{\theta_1}} \quad \text{for } m = 1,\ldots,M. \tag{9.16}$$

The other elements of the separation system are initialized to $\mathbf{w}_{nn} = \boldsymbol{\delta}_{L/2}$ for $n = 2,\ldots,M$ and $\mathbf{w}_{nm} = \mathbf{0}$ for $n = 2,\ldots,M; m = 1,\ldots,M; n \neq m$. In terms of the geometric constraint in (9.8), this corresponds to

$$\mathbf{W}(0)\mathbf{D} = \mathbf{G}_1. \tag{9.17}$$

A comment about the update in (6.54) for the initialization above is in order. We set $M = 4$ in the following. Let us define the subscript sets

$$I_1 \triangleq \{(1,2),(1,3),(1,4)\} \quad \text{and} \quad I_2 \triangleq \{(2,1),(3,1),(4,1)\}. \tag{9.18}$$

An examination of the update (6.54) shows that the updates $\boldsymbol{\Delta}\mathbf{w}_{nm}$ for $(n,m) \in I_2$ are proportional (for the convolution product) to $\mathbf{w}_{11}$. Similarly, $\boldsymbol{\Delta}\mathbf{w}_{nm}$ for $(n,m) \in I_1$ is proportional to $\mathbf{w}_{nn}$ for $n = 2,\ldots,M$. Since $\mathbf{w}_{11}(0) = \mathbf{w}_{nn}(0)/4$ for $n = 2,\ldots,M$, the initialization in (9.16) leads to smaller updates $\boldsymbol{\Delta}\mathbf{w}_{nm}$ for $(n,m) \in I_2$ than for $(n,m) \in I_1$. To compensate for this effect, we pre-multiply $\mathbf{w}_{nm}$ by 4 for $(n,m) \in I_2$, that is, we replace (6.54) with

$$\boldsymbol{\Delta}\mathbf{w}_{nm} = -4\mu \sum_{k=1}^{K} \sum_{p\in\mathcal{K}_n} \mathbf{P}_{[d,L+d-1]}^{L\times 2L-1} \left(\mathbf{w}_{pm} * \mathbf{r}_{\mathbf{y}_n\mathbf{y}_1}^{(d)}(kL)\right) / \tilde{r}_{y_ny_n,0}(kL) \tag{9.19}$$

for all $(n,m) \in I_2$.

The result for the initialization in (9.16) is shown in Table 9.2. We can see that the start-up performance $Q_{[0,3]} = 3.5$ dB is worse than the baseline. Moreover, the separation system completely breaks down in the course of the adaptation ($Q_{[3,10]} = 0.4$ dB). We observed that the signals coming from both interferer *and* target directions are reduced at the output y_1. This may be explained by the conjunction of two factors. Firstly, we have more microphones than sources. This results in additional degrees of freedom that are not controlled by the separation criterion. By contrast, if we had only two microphones ($M = N = 2$), no more than $M - 1 = 1$ spatial zero could be set and both source signals could not be canceled simultaneously. Secondly,

Table 9.2. Separation performance for geometric initializations. The results are obtained with the four-element compact array mounted in the rear-view mirror and partial BSS (PBSS)

Initial	causal (baseline) (dB)	constraint $\mathbf{G}_0$ (9.15) (dB)	constraint $\mathbf{G}_1$ (9.16) and (9.19) (dB)	constraint $\mathbf{G}_1'$ (9.21) and (9.19) (dB)
$Q_{[0,3]}$	4.0	7.2	3.5	4.3
$Q_{[3,10]}$	10.2	11.2	0.4	6.3

initializing to delay-and-sum beamformers requires an acausal separation system, and thus both source signals may be canceled at the output y_1. Therefore, the additional degrees of freedom may lead to a spurious minimum of the cost function where the two speech source signals are canceled at the output y_1.

Constraint $\mathbf{G}_1'$

This target cancelation problem may be somewhat mitigated if, in addition to setting $\mathbf{W}_1(0)$ with a unit response in the direction θ_1, $\mathbf{W}_n(0)$ for $n = 2, \ldots, M$ have a zero response in θ_1. This corresponds to the geometric constraint $\mathbf{W}(0)\mathbf{D} = \mathbf{G}_1'$ with:

$$\mathbf{G}_1' \triangleq \begin{bmatrix} \delta_{L/2} & \diamond \\ 0 & \diamond \\ 0 & \diamond \\ 0 & \diamond \end{bmatrix}. \tag{9.20}$$

The constraint $\mathbf{W}(0)\mathbf{D} = \mathbf{G}_1'$ is satisfied if the components of $\mathbf{W}(0)$ are set to

$$\begin{bmatrix} w_{11}(0) & w_{12}(0) & w_{13}(0) & w_{14}(0) \\ w_{21}(0) & w_{22}(0) & w_{23}(0) & w_{24}(0) \\ w_{31}(0) & w_{32}(0) & w_{33}(0) & w_{34}(0) \\ w_{41}(0) & w_{42}(0) & w_{43}(0) & w_{44}(0) \end{bmatrix} = \begin{bmatrix} \frac{1}{4}\delta_{\frac{L}{2}} & \frac{1}{4}\delta_{\frac{L}{2}-\tau_{\theta_1}} & \frac{1}{4}\delta_{\frac{L}{2}-2\tau_{\theta_1}} & \frac{1}{4}\delta_{\frac{L}{2}-3\tau_{\theta_1}} \\ -\delta_{\frac{L}{2}+\tau_{\theta_1}} & \delta_{\frac{L}{2}} & 0 & 0 \\ -\delta_{\frac{L}{2}+2\tau_{\theta_1}} & 0 & \delta_{\frac{L}{2}} & 0 \\ -\delta_{\frac{L}{2}+3\tau_{\theta_1}} & 0 & 0 & \delta_{\frac{L}{2}} \end{bmatrix}. \tag{9.21}$$

However, as can be seen in the last column of Table 9.2, the performance is still inferior to that of the baseline. This is also due to a target cancelation problem at the output y_1 (the target signal level is reduced by about 2.8 dB in average over the time interval [3, 10] seconds.) Maintaining the constraint of unit response in the direction θ_1 during the whole adaptation process may prevent convergence to a spurious minimum and may be realized by regularizing the separation update with a soft geometric constraint.

9.2.4 Geometric Information as a Soft Constraint

The deviation of the separation system from the geometric constraint $\mathbf{WD} = \mathbf{G}$ may be measured by the following least-square criterion:

$$J_{\text{geo}}(\mathbf{W}) \triangleq \|\mathbf{WD} - \mathbf{G}\|^2. \tag{9.22}$$

The norm is given by $\|\mathbf{G}\|^2 = \text{tr}\left(\mathbf{G}^{\mathsf{T}}\mathbf{G}\right)$. The total cost function is then defined as

$$J_{\text{total}}(\mathbf{W}) \triangleq J_{\text{BSS}}(\mathbf{W}) + \lambda J_{\text{geo}}(\mathbf{W}). \tag{9.23}$$

The parameter $\lambda > 0$ is called "geometric weight" because it controls the weight of the geometric criterion J_{geo} relative to the separation criterion J_{BSS}.

If we set $\lambda = 0$, the geometric prior information is only used at the initialization.[4] By varying the parameter λ, we may easily control the strictness of the geometric constraint. Roughly speaking, this strictness should be high if the geometric prior information matches well with the current environment and should be smaller if the geometric prior information is less reliable, for example, in a reverberant environment. Hence, an appropriate value of λ should be determined experimentally.

The parameter update $\Delta\mathbf{W}$ for minimizing J_{total} in (9.23) can be written as:

$$\Delta\mathbf{W} = \Delta\mathbf{W}^{(\text{BSS})} + \Delta\mathbf{W}^{(\text{geo})}, \tag{9.24}$$

where $\Delta\mathbf{W}^{(\text{BSS})}$ is given by (6.54). The term $\Delta\mathbf{W}^{(\text{geo})}$ is defined using the gradient of $J_{\text{geo}}(\mathbf{W})$ as follows:

$$\Delta\mathbf{W}^{(\text{geo})} \triangleq -\frac{\lambda}{2}\frac{\partial J_{\text{geo}}(\mathbf{W})}{\partial\mathbf{W}}. \tag{9.25}$$

Note that λ also acts as a step-size for the gradient descent of the cost function J_{geo}. The update $\Delta\mathbf{W}^{(\text{geo})}$ can be compactly written as

$$\Delta\mathbf{W}^{(\text{geo})} = S\left((\mathbf{W}\mathbf{D} - \mathbf{G})\mathbf{D}^{\mathrm{T}}\right). \tag{9.26}$$

The function S transforms a general matrix into a block Sylvester matrix by summing the redundant terms of Sylvester matrices (see Sect. 6.1.1). The geometric update (9.25) can be written for each $\mathbf{W}_m$ as

$$\Delta\mathbf{W}_m^{(\text{geo})} = -\frac{\lambda}{2}\sum_{\substack{n=1 \\ \mathbf{g}_{mn}\neq\diamond}}^{N} S\left((\mathbf{W}_m\mathbf{d}(\theta_n) - \mathbf{g}_{mn})\mathbf{d}^{\mathrm{T}}(\theta_n)\right). \tag{9.27}$$

In (9.27), the elements $\mathbf{g}_{mn}$ of $\mathbf{G}$ that are equal to $\diamond$ are not included in the sum simply because they correspond to an unconstrained response. For instance, the constraint $\mathbf{G}_1$ leads to an update $\Delta\mathbf{W}^{(\text{geo})}$ that contains only zeros except the first block-row $\Delta\mathbf{W}_1^{(\text{geo})}$, which is given by

$$\Delta\mathbf{W}_1^{(\text{geo})} = -\lambda\frac{\partial\|\mathbf{W}_1\mathbf{d}(\theta_1) - \boldsymbol{\delta}_{L/2}\|^2}{\partial\mathbf{W}_1} = S\left((\mathbf{W}_1\mathbf{d}(\theta_1) - \boldsymbol{\delta}_{L/2})\boldsymbol{\delta}_{L/2}^{\mathrm{T}}\right). \tag{9.28}$$

Calibrating the geometric weight λ for the car interior

First, we examine "how much" geometric prior information is useful for the car acoustical environment with the constraints $\mathbf{G}_0$ and $\mathbf{G}_1$. Artificial heads are used as speech sources. To this end, we vary the geometric weight λ in (9.25). The results for the geometric constraints $\mathbf{G}_0$ and $\mathbf{G}_1$ in (9.13) are shown in Fig. 9.1. In the case of the constraint $\mathbf{G}_0$, the performance Q is a

---

[4] $\mathbf{W}$ is initialized so that $J_{\text{geo}}(\mathbf{W}(0)) = 0$, as in Sect. 9.2.3.

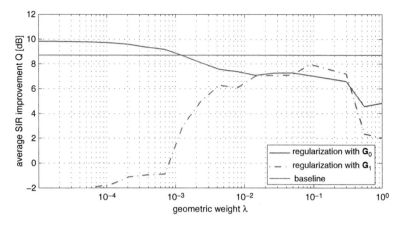

Fig. 9.1. Average SIR improvement as a function of the geometric weight λ. The performance measure Q is defined as the SIR improvement averaged over the whole signal: $Q \triangleq \mathrm{IR}_{[0,10]}/\mathrm{SR}_{[0,10]}$. ($\mathrm{IR}_{[0,10]}$ and $\mathrm{SR}_{[0,10]}$ are defined in Table 2.1.) The source signals are emitted by artificial heads on the driver and codriver seats and last 10 s. The baseline performance obtained with the causal separation system is $Q = 8.7$ dB

decreasing function of λ. This shows that maintaining the constraint $\mathbf{G}_0$ does not improve the performance. This may be explained by the poor performance of data-independent beamforming, even in the weakly reverberant car cabin. It seems better to let BSS fully control the separation system during the adaptation. Here, the simple initialization of the separation system in (9.15) leads to the best performance.

On the other hand, the constraint $\mathbf{G}_1$ yields a poor performance for $\lambda < 10^{-3}$. This reflects the result obtained by initializing the separation system with (9.16) in Sect. 9.2.3, that is, the convergence to a spurious minimum. As can be seen in Fig. 9.1, Q attains its maximum performance around $\lambda \approx 0.1$. There, the performance approximately equals the baseline. This shows that maintaining the constraint $\mathbf{G}_1$ during the whole adaptation process prevents convergence to a spurious minimum. For $\lambda > 10^{(-0.5)}$, the update $\mathbf{\Delta W}_1^{(\mathrm{geo})}$ in (9.28) is too large, which leads to instability.

Results on real speakers and improvements

Let us now focus on the constraint $\mathbf{G}_1$ and set $\lambda = 0.1$. Applying this constraint to the separation of real speakers, we obtain the performance reported in the third column of Table 9.3. The spurious minimum problem is solved but the performance $Q_{[3,10]} = 9.0$ dB is not better than the baseline. There is still room for improvement.

A useful observation is that the coefficients $w_{1m,k}$ in $\mathbf{W}_1$ shape the spatial response differently depending on their position with respect to the delay

Upper half plane

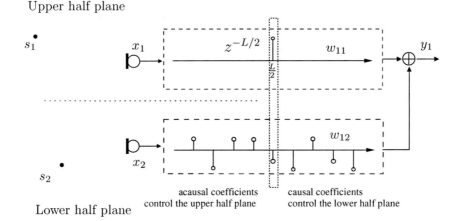

Lower half plane

Fig. 9.2. Depending on their position relative to $L/2$, the filter coefficients influence different parts of the spatial response

Table 9.3. Performance of NG-SOS-BSS algorithms with geometric constraint applied as soft constraint. The results are obtained with the four-element compact array mounted in the rear-view mirror and PBSS

Soft constraint	causal (baseline) (dB)	constraint $\mathbf{G}_1$ (9.28) with $\lambda = 0.1$ (dB)	constraint $\mathbf{G}_1$ (9.28), (9.29) with $\lambda = 0.1$ (dB)
$Q_{[0,3]}$	4.0	5.2	10.0
$Q_{[3,10]}$	10.2	9.0	11.3

(or acausal length) d. Let us set $d = L/2$ and assume a far- and free-field acoustic propagation. As shown in Fig. 9.2, the coefficients $w_{1m,k}$ for $k < L/2$ may enhance or reduce the sources in the upper half plane. In other words, these coefficients are responsible for enhancing or canceling the target source signal. On the other hand, the coefficients $w_{1m,k}$ for $k > L/2$ shape the spatial response in the lower half plane, where the interferer is located. Now, we want to control the spatial response to prevent the convergence toward a *target*-canceling spurious minimum. According to Fig. 9.2, this may be done by constraining the coefficients $w_{1m,k}$ for $k \leq L/2$ only. Hence, we set the geometric update to zero for the other coefficients:

$$\Delta w_{1m,k}^{(geo)} = 0 \quad \text{for } k = L/2 + 1, \ldots, L. \tag{9.29}$$

Then the coefficients $w_{1m,k}$ for $k > L/2$ are *not* influenced by the geometric prior information. They are entirely controlled by the BSS criterion.

The result, which is given in Table 9.3, shows a significant increase in the separation performance against the baseline. In particular, in terms of

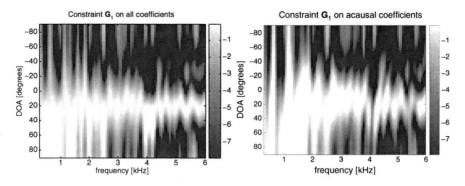

Fig. 9.3. Beampattern of $\mathbf{W}_1$ after one pass in the 10-s signals. The NG-SOS-BSS update was regularized with the geometric constraint $\mathbf{G}_1$. *Left*: The soft constraint is applied to all coefficients $w_{1m,k}$, $k = 0, \ldots, L$. *Right*: The soft constraint is applied to the acausal coefficients $w_{1m,k}$, $k \leq L/2$, according to (9.29)

start-up performance $Q_{[0,3]}$, the algorithm clearly outperforms the baseline and shows that a careful use of geometric prior information may lead to a substantial improvement of the performance. Note that the constraint $\mathbf{G}_1$ is not maintained during the adaptation and the question arises of how zeroing the updates in (9.29) affects the geometric constraint. To answer this question, the beampatterns of $\mathbf{W}_1$ with and without zeroing are shown in Fig. 9.3. It may be seen that the constraint of unit response in the target direction, though not strictly, is nevertheless sufficiently maintained. Therefore, the problem of target-canceling spurious minimum is efficiently prevented.

9.2.5 Geometric Information as a Preprocessing

If we maintain the geometric constraint strictly, the minimization of $J_{\mathrm{BSS}}(\mathbf{W})$ becomes a constrained problem:

$$\min_{\mathbf{W}} J_{\mathrm{BSS}}(\mathbf{W}) \quad \text{s.t.} \quad \mathbf{WD} = \mathbf{G}. \tag{9.30}$$

An analogous situation appears in the design of LCMV beamformers. In (9.30) and in the LCMV problem (3.7), the filter coefficients are adjusted to minimize a cost function and to fulfill a linear constraint. As shown in Chap. 3, the generalized sidelobe canceler (GSC) transforms the LCMV constrained minimization problem into an unconstrained problem using a spatial preprocessor. Likewise, this section presents examples of spatial preprocessors to solve the problem (9.30) for some particular constraints $\mathbf{G}$. Since $\mathbf{W}$ denotes the entire separation system, we denote the filters in the BSS adaptive part by $\mathbf{W}'$, as depicted in Fig. 9.4.

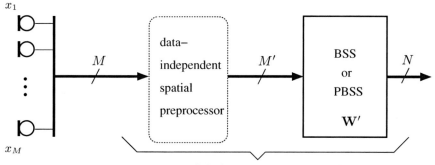

Fig. 9.4. Two-stage separation system consisting of a data-independent spatial pre-processor and an adaptive source separation block

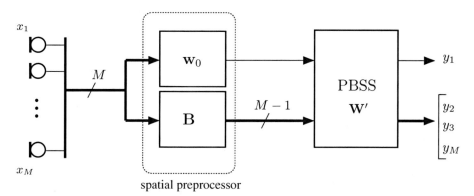

Fig. 9.5. The spatial preprocessor for the constraint $\mathbf{G}_1$ leads to the "generalized sidelobe decorrelator" (GSD)

The generalized sidelobe decorrelator

First, we examine the case of the constraint $\mathbf{G}_1$, which restricts $\mathbf{W}_1$ to have a unit response in the direction θ_1. We directly deduce from the analogy with the LCMV beamformer that the spatial preprocessor from the GSC may be used to transform the constrained problem (9.30) for $\mathbf{G} = \mathbf{G}_1$ in an unconstrained one. The resulting structure is depicted in Fig. 9.5. We just need to replace the adaptive interference canceler of a GSC with the PBSS algorithm. This structure has been called "generalized sidelobe decorrelator" (GSD) because of the similarity with the GSC, and was proposed by Fancourt et al. [35]. It should be mentioned that we use $M - 1$ interferer output signals in the structure shown in Fig. 9.5. By contrast, the original GSD proposed by Fancourt et al. includes only one output signal for the interferer [35].

Note that the constraint $\mathbf{WD} = \mathbf{G}_1$ is satisfied strictly only if the component $\mathbf{w}'_{11}$ of $\mathbf{W}'$ is fixed to $\mathbf{w}'_{11} = \boldsymbol{\delta}_{L/2}$. The mixing that is to be separated by the adaptive PBSS block results from the combination of acoustic impulse responses and the spatial preprocessing. This mixing is not necessarily causal. Hence, similarly to the GSC, the filters $\mathbf{w}'_{mm}$ in the adaptive block are initialized with $\mathbf{w}'_{mm} = \boldsymbol{\delta}_{L/2}$ for $m = 1, \ldots, M$.

The constraint $\mathbf{G}_0$

The simple combination of GSC and blind source separation in the GSD is possible because the spatial preprocessor provides signals for all sources: the spatial preprocessor provides a target reference signal and interference reference signals. Let us consider, by contrast, a constraint of zero response such as $\mathbf{W}_1 \mathbf{d}(\theta_2) = \mathbf{0}$. This constraint defines a linear subspace $\{\mathbf{W}_1 \text{ s.t. } \mathbf{W}_1 \mathbf{d}(\theta_2) = \mathbf{0}\}$. As in the case of the GSC, we may derive a spatial preprocessor with the basis vectors of this subspace. All these vectors have a zero response in the direction θ_2. This implies that the source signal $s_2(p)$ is canceled by the preprocessor. Consequently, $s_2(p)$ cannot be recovered by the separation system. Therefore, $\mathbf{G}_0$ may not be implemented as a hard constraint.

Multiple constraints

The spatial preprocessor of the GSD relies on the prior knowledge of θ_1 only. However, the prior information on θ_1 *and* θ_2 may be exploited if we use the null beamformer as derived in Sect. B.2. This data-independent LCMV null beamformer, which has a unit response in the direction θ_1 and a zero response in the direction θ_2, is denoted by $\mathbf{w}_{[\theta_1, \theta_2]}$. Using its counterpart $\mathbf{w}_{[\theta_2, \theta_1]}$, which has a unit response in the direction θ_2 and a zero response in the direction θ_1, we may implement the spatial preprocessor as depicted in Fig. 9.6. Let us define the constraint $\mathbf{G}_2$ as follows:

$$\mathbf{G}_2 \triangleq \begin{bmatrix} \boldsymbol{\delta}_{L/2} & \diamond \\ \diamond & \boldsymbol{\delta}_{L/2} \end{bmatrix}. \tag{9.31}$$

If we hold the diagonal filters of the BSS block in Fig. 9.6 to $\mathbf{w}'_{nn} = \boldsymbol{\delta}_{L/2}$ for $n = 1, 2$, this preprocessing transforms the constrained problem (9.30) for $\mathbf{G} = \mathbf{G}_2$ into an unconstrained one.

The results for hard geometric constraints $\mathbf{G}_1$ and $\mathbf{G}_2$ are reported in Table 9.4. For the GSD (i.e., the constraint $\mathbf{G}_1$), there is no noticeable improvement over the baseline. This may be explained by the fact that an acausal separation system $\mathbf{W}'$ must be used, which is unfavorable as shown in Sect. 9.2.1. On the other hand, there is an improvement for the constraint $\mathbf{G}_2$, in particular for the start-up performance $Q_{[0,3]}$. This may be explained by the reduction of the crosstalk that is performed by the spatial preprocessor. As shown in Sect. 7.2, a lower crosstalk level at the input is beneficial (although the separation system $\mathbf{W}'$ is acausal).

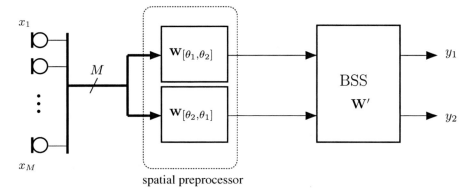

Fig. 9.6. Spatial preprocessor for the constraint $\mathbf{G}_2$. $\mathbf{w}_{[\theta_1,\theta_2]}$ denotes the data-independent null beamformer which has a unit response in the direction θ_1 and a zero response in the direction θ_2

Table 9.4. Performance with geometric prior information applied as hard constraints. These results are obtained with the four-element compact array mounted in the rear-view mirror and the self-closed NG-SOS-BSS updates. For the constraints $\mathbf{G}_1$ and $\mathbf{G}_2$, the separation system was initialized as in (6.69) for $d = L/2$

Hard constraint	causal (baseline) (dB)	constraint $\mathbf{G}_1$ (GSD, Fig. 9.5) (dB)	constraint $\mathbf{G}_2$ (Fig. 9.6) (dB)
$Q_{[0,3]}$	4.0	5.8	9.6
$Q_{[3,10]}$	10.2	9.4	11.4

9.3 Combining SOS-BSS and the Power Criterion

In the previous section, we have derived techniques to integrate geometric prior information into time-domain BSS. In this section, we propose to adapt the separation system with the SOS-BSS criterion and the LMS power criterion in adaptive LCMV beamforming.

Let us assume that no permutation occurs and consider that the signal $y_1(p)$ is an estimate of the desired signal while the other outputs may be considered as interferer references. This assumption holds if the separation system is causal, for example. We may consider BSS as a method that provides interferer references in the output $y_2(p), \ldots, y_M(p)$, and place an adaptive interference canceler after the BSS block as shown in Fig. 9.7. Positive effects may result from this combination of SOS-BSS and the LMS power criterion.

- A first effect is the improved interference cancelation during target silences. The NG-SOS-BSS step-size being relatively small to prevent instability, the convergence of the BSS block is slower than the convergence of the

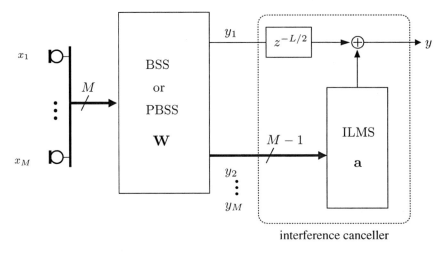

Fig. 9.7. Combination of BSS and LMS power criterion. For the BSS/PBSS stage, we may use a causal separation system

interference canceler. This slow convergence may be compensated by the interference canceler, which has fast tracking capabilities.

- Regarding the target signal cancelation problem, a second positive effect may result from the combination of SOS-BSS with an interference canceler. Including a BSS block as in Fig. 9.7 may provide interferer references that may be sufficiently free of leakage to prevent the cancelation of the target signal at the output of the interference canceler.

It should be noted that the relatively slow convergence of NG-SOS-BSS algorithms does not allow for a rapid suppression of the desired signal leakage in the interferer reference. This negative effect may be balanced if the interference canceler is adapted by the ILMS algorithm (4.13).

As shown in Chap. 4, ILMS adapts slower when the target signal is active. This property alone is generally not sufficient to prevent the target signal cancelation after convergence (as in the case of the four-element compact array mounted in the rear-view mirror, see Fig. 4.2). However, this property may make the slow convergence of the BSS block still sufficient to prevent desired signal cancelation.

In analogy with the generalized sidelobe canceler (GSC), the interference canceler is implemented with filters that have causal and acausal parts of equal length, hence the $L/2$-taps delay. Note that the interferer references may be correlated to each other. Thus, PBSS may be used instead of BSS if $M > N$. The structure shown in Fig. 9.7 is denoted PBSS-ILMS (or BSS-ILMS if $M = N$). A similar structure was proposed in the subband domain by Low et al. [67].

The results for both microphone setups are reported in Table 9.5. In the case of the four-element compact array mounted in the rear-view mirror, the

Table 9.5. Performance of NG-SOS-BSS with the LMS power criterion for the four-element compact array mounted in the rear-view mirror and the two-element distributed array mounted on the car ceiling

	compact array		distributed array		
BSS	causal	PBSS-ILMS	causal	ILMS alone	BSS-ILMS
and LMS	(baseline)	Fig. 9.7	(baseline)	(Sect. 4.5.2)	Fig. 9.7
	(dB)	(dB)	(dB)	(dB)	(dB)
$Q_{[0,3]}$	4.0	6.2	6.8	13.0	12.9
$Q_{[3,10]}$	10.2	8.0	11.4	9.8	11.3

start-up performance $Q_{[0,3]}$ is slightly improved relative to the PBSS baseline. This result reflects the fast convergence of the ILMS-adapted interference canceler, which provides a rather high interference signal suppression (about 11.3 dB). However, the interference canceler also reduces the target signal, which leaks into the interferer references. The target signal suppression averaged on the first three seconds ($SR_{[0,3]}$) is about 5.0 dB. The suppression of the target signal is mitigated in the course of the adaptation because the PBSS block eventually converges and reduces the target leakage. In spite of this, the long-term performance $Q_{[3,10]}$ is inferior to the baseline.

In the case of the two-element distributed array mounted on the car ceiling, no significant target signal cancelation occurs. This may be explained from the results in Chap. 4: The ILMS-adapted interference canceler does not lead to severe target signal cancelation. However, it increases the start-up performance $Q_{[0,3]}^d$. The long-term performance $Q_{[3,10]}^d$ remains approximately unchanged. As shown in Table 9.5, BSS-ILMS combines the best scores $Q_{[0,3]}^d$ and $Q_{[3,10]}^d$ of ILMS and BSS, respectively.

9.4 Combining SOS-BSS with Geometric Prior Information and the Power Criterion

In this section we propose a GSC-based structure that combines SOS-BSS with an adaptive interference canceler. That is, this structure combines SOS-BSS with the LMS power criterion as in Sect. 9.3, and exploits geometric prior information as in Sect. 9.2.

Similarly to the RGSC[5] (robust GSC) proposed by Hoshuyama et al., we consider an adaptive blocking matrix [53]. In the RGSC proposed by Hoshuyama et al., the adaptation of the blocking matrix and of the interference canceler should only occur for a dominant target signal and dominant interferer signal, respectively. Using the partial BSS approach (PBSS), we may design a structure where both the blocking matrix and the interference canceler are continuously adapted. The proposed structure is depicted in Fig. 9.8.

---

[5] See Sect. C.1 for more details.

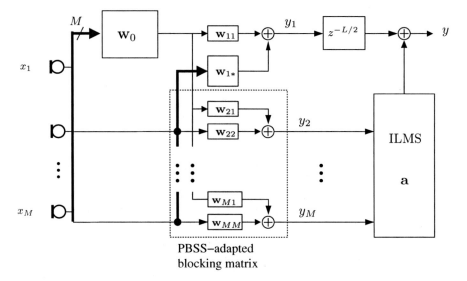

Fig. 9.8. The blind GSC (BGSC) is a combination of SOS-BSS with geometric prior information and the LMS power criterion. The delay-and-sum fixed beamformer is denoted by $\mathbf{w}_0$

- Considering the high level of cross-talk at the system input, PBSS is suitable for the adaptation of the blocking matrix. Since the outputs of the blocking matrix are allowed to be correlated to each other, a partial separation is sufficient.
- The adaptation of the blocking matrix limits the leakage of the target signal, but the interferer references may not be perfectly free of target signal. Hence, as in Sect. 9.3, the adaptation of the interference canceler may be carried out with the ILMS algorithm (4.13) rather than with the standard NLMS (4.7).

In analogy with the RGSC, the blocking-matrix filters $\mathbf{w}_{nm}$ are initialized as[6]

$$\begin{bmatrix} \mathbf{w}_{11}(0) & \mathbf{w}_{12}(0) & \mathbf{w}_{13}(0) & \mathbf{w}_{14}(0) \\ \mathbf{w}_{21}(0) & \mathbf{w}_{22}(0) & \mathbf{w}_{23}(0) & \mathbf{w}_{24}(0) \\ \mathbf{w}_{31}(0) & \mathbf{w}_{32}(0) & \mathbf{w}_{33}(0) & \mathbf{w}_{34}(0) \\ \mathbf{w}_{41}(0) & \mathbf{w}_{42}(0) & \mathbf{w}_{43}(0) & \mathbf{w}_{44}(0) \end{bmatrix} = \begin{bmatrix} \delta_{L/2} & \mathbf{0} & \mathbf{0} & \mathbf{0} \\ -\delta_{L/2} & \delta_{L/2} & \mathbf{0} & \mathbf{0} \\ -\delta_{L/2} & \mathbf{0} & \delta_{L/2} & \mathbf{0} \\ -\delta_{L/2} & \mathbf{0} & \mathbf{0} & \delta_{L/2} \end{bmatrix}. \quad (9.32)$$

This initialization sets the spatial response of the blocking matrix to zero in the direction of the target signal. The PBSS-adapted blocking matrix in Fig. 9.8 is similar to the adaptive blocking matrix of the RGSC. However, they have different dimensions. Whereas the RGSC structure according to

---

[6] It is assumed that the array is steered to θ_1 so that the target signal reaches the microphones synchronously.

Table 9.6. Performance of SOS-BSS combined with geometric prior information and the power criterion

BSS, geometric info. and power criterion	causal PBSS (baseline) (dB)	blind GSC (BGSC) Fig. 9.8 (dB)
$Q_{[0,3]}$	4.0	15.2
$Q_{[3,10]}$	10.2	11.9

Hoshuyama et al. has M outputs, the PBSS-adapted blocking matrix has $M-1$ outputs, like the original GSC [42].

As opposed to other GSCs, the structure that we propose does not require an external adaptation control mechanism. Hence, we refer to it as the blind GSC (BGSC). Let us mention that the target reference signal $y_1(p)$ does not only result from the delay-and-sum fixed beamformer $\mathbf{w}_0$ but also from an adaptive part $\mathbf{W}_1$. Since the filter coefficients in $\mathbf{W}_1$ are not subject to any spatial constraint, the structure in Fig. 9.8 does not belong to the class of the linearly constrained beamformers (in contrast to the GSC).

The experimental result is reported in Table 9.6. It shows a significant improvement of the start-up performance, which is due to the high interference cancelation provided by the ILMS-adapted interference canceler. A closer examination of the output signals reveals that the target signal level is also reduced, because the PBSS-adapted blocking matrix converge slower than the interference canceler. Nevertheless, this reduction is much smaller than in the case of PBSS-ILMS of Fig. 9.7. We have $SR_{[0,3]} = 1.7$ dB for the BGSC, which may be compared to $SR_{[0,3]} = 5.0$ dB in the case of PBSS-ILMS. Even though the target signal level reduction remains higher than for PBSS, it diminishes steadily due to the adaptation of the blocking matrix (we have $SR_{[3,10]} = 0.6$ dB). It appears that the output signals of the PBSS-adapted blocking matrix is sufficiently free of target components to allow a continuous adaptation of the interference canceler, also during target source activity. The cancelation of the interference signal is also higher than that of PBSS baseline.

9.5 Experimental Results on Automatic Speech Recognition

This section evaluates the proposed algorithms as an acoustic front-end for a speech recognizer. The goal is to provide a robust speech recognition for the driver when the codriver speaks simultaneously.

Testing conditions

We used the speaker-independent DaimlerChrysler recognizer[7] for semi-continuous speech. The test data consist of speech from 28 speakers (14 males,

---

[7] Let us briefly describe the DaimlerChrysler recognizer. Using a linear discriminant analysis (LDA), this HMM recognizer extracts cepstral features from nine

14 females), which was played back by an artificial head. This artificial head was placed first on the driver seat, and then on the codriver seat in a car. For each speaker, 40 occurrences were played back. The microphone signals were recorded with sampling rate $f_s = 16$ kHz. These test speech signals were taken from the TIDigits corpus and each occurrence consists of up to 7 digits. Accordingly, the recognizer lexicon consists of digits only ("1"–"10", "oh" and "zero"). The input signals are generated as the sum of the separately recorded driver and codriver signals. This simulates the situation which is depicted in Fig. 9.9a. These input signals exhibit a rather high degree of overlap between the target and the interferer (double-talk situation). Additionally, test data were recorded for the back seat passenger, as depicted in Fig. 9.9b. Furthermore, road noise was recorded and added to the speech signals. Depending on the speaker loudness, the SNR varies between 10 and 20 dB.

Performance measures

We denote the number of words uttered by the target speaker (reference words) by N_{target}. The number of deletions, i.e., missed reference words, is denoted by $N_{\text{deletions}}$. The number of substitutions, i.e., reference words substituted by others, is denoted by $N_{\text{substitutions}}$. The number of insertions, i.e., wrong words inserted between correctly recognized reference words, is denoted by $N_{\text{insertions}}$. We consider two performance measures: the word accuracy and the word error rate (WER). The word accuracy is the percentage of recognized reference words, that is,

$$\text{word accuracy} = 100 \left(1 - \frac{N_{\text{deletions}} + N_{\text{substitutions}}}{N_{\text{target}}} \right). \tag{9.33}$$

The word accuracy depends on the number of deletions and substitutions but does not include the insertions. The WER may be defined as follows:

$$\text{WER} \triangleq 100 \frac{N_{\text{deletions}} + N_{\text{substitutions}} + N_{\text{insertions}}}{N_{\text{target}}}. \tag{9.34}$$

In contrast to the word accuracy, the WER takes the insertions into account. In the context of competing speakers, many insertions may lead to both high word accuracy and high WER.

---

successive frames of the acoustic input signals. A cepstral mean normalization is applied to the feature vectors to remove influences of speaker, microphone, and room acoustics. Training data were recorded in the car in various noisy conditions (engine, wind noise) and consist of 300,000 utterances from about 1,000 speakers. These data were used to estimate the HMM transition and emission parameters with the Baum-Welch algorithm. The recognizer also includes a one-channel noise reduction algorithm (spectral subtraction), which we disabled for noise-free input signals.

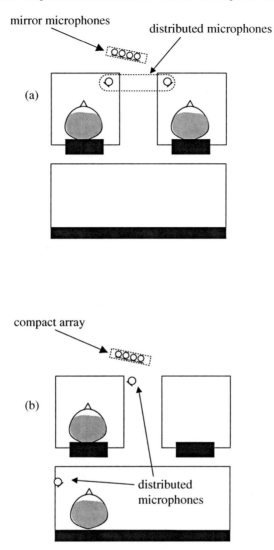

Fig. 9.9. Experimental layouts for the speech recognition tests. (a) Separation of driver and codriver; (b) separation of driver and back-seat passenger

Results

The word accuracy and the relative improvement of the WER are reported for the four-element compact array mounted in the rear-view mirror and the two-element distributed array mounted on the car ceiling in Tables 9.7 and 9.8. The

Table 9.7. Word accuracy and word error rates (WER) for the four-element compact array mounted in the rear-view mirror. The interferer is the codriver speech. (a) All algorithms are compared without road noise. (b) The best algorithms were also tested in noisy conditions

	word accuracy	WER		word accuracy	WER
microphone x_1	72.6	80.0	microphone x_1	69.8	80.7
delay-and-sum	75.1	75.1	delay-and-sum	72.3	76.9
PBSS-ILMS	78.9	22.8	BSS (baseline)	83.7	39.3
preprocessing $\mathbf{G_2}$	85.1	32.9	initialization $\mathbf{G_0}$	85.0	37.2
soft constraint $\mathbf{G_1}$	85.6	41.2	DTD-NLMS (RGSC)	85.3	20.2
BSS (baseline)	85.9	41.9	GSD $\mathbf{G_1}$	86.7	25.7
DTD-NLMS (RGSC)	86.8	19.9	BGSC	87.6	14.7
initialization $\mathbf{G_0}$	88.7	33.4			
GSD $\mathbf{G_1}$	89.4	29.5			
BGSC	90.0	12.0			

(a) noise-free conditions	(b) with road noise

Table 9.8. Word accuracy and word error rates (WER) for the two-element distributed array mounted on the car ceiling. The interferer is the codriver speech. (a) All algorithms are compared without road noise and (b) in noisy conditions

	word accuracy	WER		word accuracy	WER
microphone x_1	76.7	75.5	microphone x_1	76.1	74.2
ILMS	84.2	22.1	ILMS	85.7	18.0
DTD-NLMS	87.2	16.0	DTD-NLMS	86.4	19.3
BSS-ILMS	87.7	15.6	BSS (baseline)	88.2	16.1
BSS (baseline)	89.2	20.9	BSS-ILMS	88.4	13.8

(a) noise-free conditions	(b) with road noise

performance of the DTD-NLMS algorithms is also given. For this algorithm, the target activity detection was done by comparing the short-term energy of the driver signal alone with a fixed threshold.

We observe that the word accuracy is not always in agreement with the SIR performances given in Tables 9.2–9.6. In the case of the four-element compact array mounted in the rear-view mirror, all algorithms lead to an improvement of WER, especially those with an interference canceler (BSS-ILMS, DTD-NLMS, and BGSC). However, not all are able to improve the word accuracy. As may be expected from the result in Table 9.5, the BSS-ILMS algorithm reduces the target signal, which decreases the word accuracy. On the other hand, algorithms that do not include an interference canceler may increase the word accuracy but lead to an inferior improvement of the WER. That is, these algorithms are less efficient against insertions. By contrast, BGSC leads to the best increase in the word accuracy *and* to the best improvement of the WER.

Table 9.9. Comparison of the four-element compact array mounted in the rear-view mirror (a,b) and the two-element distributed array mounted on the car ceiling (c,d) with speech from the backseat passenger as interferer

	word accuracy	WER		word accuracy	WER
microphone x_1	82.6	61.5	microphone x_1	79.2	56.1
delay-and-sum	82.8	62.0	delay-and-sum	81.3	58.4
BGSC	88.1	28.3	BGSC	83.5	27.7
(**a**) noise-free conditions			(**b**) with road noise		

	word accuracy	WER		word accuracy	WER
microphone x_1	79.7	69.8	microphone x_1	77.9	68.0
BSS-ILMS	90.4	10.8	BSS-ILMS	89.9	11.4
(**c**) noise-free conditions			(**d**) with road noise		

In the case of the two-element distributed array mounted on the car ceiling, it is remarkable that the BSS baseline algorithm outperforms the DTD-NLMS algorithm without using any double-talk detector. This may be explained by the fact that the interference signal level reduction of the DTD-NLMS algorithm relies importantly on the spectral content of the interferer, which may not be tracked in the case of frequent double-talk situation. The BSS-ILMS algorithm brings a small improvement over the BSS baseline in noisy conditions.

Additional tests

In an additional test, the interferer was placed on the back seat. The situation, which is depicted in Fig. 9.9b, is particularly challenging for the four-element compact array mounted in the rear-view mirror because the target speaker, the interferer, and the array are nearly aligned. The results are reported in Table 9.9. In spite of the difficulty, the proposed BGSC algorithm leads to an improvement of the word accuracy which is significant in noise-free conditions. As expected, the distributed microphone arrangement is more adequate and performs better in terms of WER.

9.6 Summary and Conclusion

In this chapter, we have developed techniques (1) to include geometric prior information into BSS algorithms, and (2) to combine a BSS separation system with an interference canceler.

We found that certain geometric initialization of the separation system may yield a convergence to a spurious minimum. Careful design of the geometric constraint, in particular regarding the causality of the separation system,

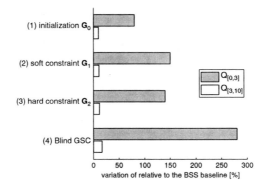

Fig. 9.10. Improvement of the SIR performance relative to the BSS baseline for combinations of NG-SOS-BSS and beamforming. (1) Geometric initialization (9.15); (2) soft constraint $\mathbf{G}_1$ with (9.28), (9.29), and $\lambda = 0.1$; (3) hard constraint $\mathbf{G}_2$ (Fig. 9.6); (4) BGSC described in Sect. 9.4

may be required. Then, as shown in Fig. 9.10, the input of geometric prior information may lead to a significant improvement of the start-up performance. However, the improvement of the performance after the initial convergence phase is limited.

The use of the ILMS interference canceler as a postprocessing deserves a special comment. This postprocessing leads to higher cancelation of both interference and desired signal. Fortunately, the desired signal cancelation may be kept moderate if geometrical prior information is adequately used, either at the physical level (BSS-ILMS with the two-element distributed array mounted on the car ceiling) or at the algorithmic level (BGSC with the four-element compact array mounted in the rear-view mirror). Then the desired signal cancelation is uncritical and the algorithm may be used as an efficient front-end in automatic speech recognition.

Using the results from Sect. 8.2 for $L = 256$, one can show that the complexity of BGSC is larger but comparable to that of the DTD-NLMS algorithm in the RGSC structure, due to the efficient implementation of the convolution in the DFT domain for the BSS part. In contrast to the RGSC, the BGSC does not require an adaptation control. Therefore, no "universal" threshold needs to be set, which makes BGSC an attractive algorithm when operating under various a priori unknown conditions.

10
Summary and Conclusions

Safety and convenience issues require hands-free speech-based human–machine interfaces to manipulate complex functionalities and devices, for example, in cars. Such interfaces severely suffer from local interferences, such as the codriver voice, which have to be suppressed. As the desired signal and the local interference have the same nature (speech), it is difficult to separate them from their temporal (or spectral) properties. However, since they are emitted from different locations, separation of the desired signal may be resolved by exploiting the spatial properties of the source signals using microphone arrays.

A particularity of the car environment is that the position of the speakers relative to the array is known a priori. This prior information may be used directly using a linearly constrained minimum variance (LCMV) beamformer. Adaptive LCMV beamformers are able to attain a high suppression of the interference signal. Unfortunately they may also cancel the desired signal if the adaptation occurs during target signal activity. This target signal cancelation problem is a central motivation for the further investigations developed in this book.

Most existing approaches to this problem are based on an all-or-nothing adaptation control that interrupts the adaptation when an estimate of the signal-to-interference ratio is below a given decision threshold. In the case of simultaneous concurrent speakers, this strategy generally yields limited interference cancelation since the adaptation is interrupted during double-talk. Moreover, the tuning of an appropriate decision threshold may be difficult. In this book, we addressed the separation problem using continuous, uninterrupted adaptive algorithms, which may be able to adapt continuously without requiring any decision threshold.

In the first part of this book (Chaps. 3 and 4), the focus was on LCMV beamforming. Building upon the widely used normalized least-mean-square (NLMS) algorithm, we devised the implicit LMS (ILMS) algorithm which implicitly includes an adaptation control and does not require any threshold.

Also, we experimentally found out that the ILMS step-size parameter automatically adjusts to the background noise level, as opposed to NLMS. This is a highly desirable feature for automotive applications. We have experimentally observed that ILMS mitigates the target signal cancelation substantially in the case of the two-element distributed array mounted on the car ceiling. However, in the case of the four-element compact array mounted in the rear-view mirror, it does not sufficiently reduce the target signal cancelation. In this case, more sophisticated blind source separation techniques (BSS) are necessary.

The second part of this book (Chaps. 5–7) was devoted to BSS techniques. While beamforming techniques are based on spatial prior information, BSS techniques exploit the statistical independence of the source signals. Based on the time-domain approach by Buchner et al. [18, 20], natural gradient second-order statistics BSS algorithms (NG-SOS-BSS) were introduced using the Sylvester representation of FIR filters. Then, these algorithms were extended along two axis:

- The natural gradient applies to square systems, where the number of microphones equals the number of sources. Comparing possible extensions of the natural gradient algorithms to nonsquare systems, we introduced the concept of "partial blind source separation" (PBSS). At a moderate computational cost, PBSS flexibly exploits all microphone signals. In addition to extracting the desired signal, PBSS provides multiple interferer references.

- The derivation of NG-SOS-BSS algorithms in Chap. 5 did not directly yield implementable coefficient updates, since the Sylvester matrix form was ignored. This problem was tackled by deriving a general convolutive formulation of the natural gradient for an arbitrary cost function J. Two types of updates arise from this analysis: the self-closed updates, which depend on a parameter d controlling the acausal length (or delay) in the separation system, and the non-self-closed updates. It appeared that any row or the Lth column may be chosen as reference to maintain the Sylvester structure, in addition to the two choices given in [5]. Our derivation provided self-closed and non-self-closed update rules for both causal and acausal separation systems. In particular, new self-closed update rules for acausal separation systems have been proposed.

In the case of NG-SOS-BSS, strict use of the derived natural gradient updates require matrix inversions. To avoid them, the NG-SOS-BSS updates are approximated and efficient learning rules have been obtained. An experimental comparison was carried out with real signals, and the self-closed updates emerged as the most robust ones for causal and acausal separation systems. The comparison with other widely used frequency-domain BSS algorithms demonstrated the good performance of the self-closed time-domain algorithm.

These extensions resulted in implementable, flexible, and efficient BSS updates. Also, an emphasis was placed on the theoretical study of the properties of BSS. Thereby, the role of the causality was evidenced at two levels:

- As illustrated in Fig. 6.2, using a causal separation system determines the ordering of the estimated source signals. That is, the so-called "permutation ambiguity" is removed. There is a price for this desirable property: A causal separation system cannot be used (1) if the sources are placed in the same half-plane, or (2) if the microphone signals are passed in a certain preprocessing such as the delay-and-sum fixed beamforming.
- Using the self-closed update for $d = 0$, the global convergence of the separation system depends on the mixing system only through the initial point ("equivariant" learning rules).

A more precise analysis of the convergence was possible only to a limited extent. The global convergence of a decorrelation algorithm was analyzed in the case of instantaneous mixtures, and this analysis served as a basis to develop a dynamic regularization scheme for SOS-BSS in the convolutive case.

By contrast, the local stability could be analyzed in the convolutive case rigorously: We have shown that the local stability conditions set an upper bound on the amount of cross-talk.

Many existing algorithms for convolutive audio BSS are derived in the frequency domain. They are based on the narrowband signal model which allows to transform a convolutive mixture in a collection of instantaneous mixtures in each frequency bin. The insights gained from the time-domain discussion could not be found out using the narrowband frequency-domain model. The narrowband frequency-domain mixture model hides the temporal aspects of the acoustic propagation. These temporal aspects, in particular the causality, shed an instructive light on the permutation problem.

The purpose of the last part of the book (Chaps. 8 and 9) was twofold: First, we wanted to give a detailed comparison of SOS-BSS and LCMV beamforming. Our second objective was to combine these two approaches efficiently. The car interior served as a privileged test environment.

It is clear that both approaches tackle the same signal separation problem using different a priori information. Technically, they are based on different cost functions and minimization algorithms with specific convergence properties. We have shown that in fact both approaches take advantage of the source silences. But in contrast to LMS-adapted LCMV beamforming, the NG-SOS-BSS step-size needs to be relatively small to prevent instability. For this reason, depending on the degree of overlap between the desired signal and the interferer, the initial convergence speed of NG-SOS-BSS may be low compared to LMS-adapted LCMV beamforming.

In Chap. 9, we presented several methods to include geometrical prior information in NG-SOS-BSS and measured the performance improvement in the car acoustic environment. The acausal self-closed update derived in Chap. 6 was used extensively since the geometrical constraint may require an acausal separation system. While the input of geometrical prior information may increase the start-up performance, we found that the performance gain after the initial convergence is limited. The use of an adaptive interference canceler as a postprocessing leads to a higher interference suppression, and also to a higher cancelation of the desired signal. Fortunately, the cancelation of the desired signal may be kept moderate by using the ILMS algorithm *and* geometrical prior information adequately, either at the physical level (BSS-ILMS with the two-element distributed array mounted on the car ceiling) or at the algorithmic level (BGSC with the four-element compact array mounted in the rear-view mirror). Then the cancelation of the desired signal is uncritical and the proposed algorithm may be used as an efficient front-end in automatic speech recognition.

Self-criticism, open issues, and future research

The present work demonstrated the feasibility of separating speech sources using continuous, uninterrupted adaptive algorithms and the benefit of combining BSS and beamforming techniques, but a number of issues remain open. Let us discuss a few of them.

- The two typical choices S_1 and S^L shown in Fig. 6.1 (the first row and the Lth column) emerged from our derivation of the natural gradient for convolutive mixtures. However, the efficient learning rules could be obtained only at the price of approximations. The meaning of these approximations was not really understood, and their impact in terms of performance was not investigated.
- The convergence properties of NG-SOS-BSS algorithm remain largely unknown. Even though the analysis of the global convergence may not be directly tackled, one may still improve on the results proposed in this book. For instance, the ILMS algorithm might be easier to analyze than the NG-SOS-BSS, since it is closely related to the well-known NLMS, and its study might improve our understanding of the NG-SOS-BSS due to the tight connections between ILMS and the sample-wise NG-SOS-BSS algorithm.

  The regularization proposed in (7.26) was motivated by the convergence analysis in Sect. 7.1.2 but it was not satisfactory in the sense that the choice of the step-size is still relatively sensitive. More robust techniques might be developed. The convergence analysis in Sect. 7.1.2 may still serve as a basis, but the ILMS algorithm, which has known stability conditions, may also provide a support.

Also, on the technical side, other options may have been followed.

- The robustness of the separation methods to background noise was assessed experimentally, but it might be worth developing specific countermeasures.

- We limited the geometric prior information to direction of arrivals (DOA) using the simple far- and free-field acoustic propagation model. A finer model of the acoustic environment might improve the performance, for example, using the norm of the acoustic channels measured a priori as proposed in [2].

A

Experimental Setups

There is still no universal agreement on the optimum placement of the microphones for the speech input in cars. Our experiments have been carried out in a Mercedes S320 vehicle with two different microphone arrays. An overview is shown in Fig. A.1.

A.1 The Four-Element Compact Array Mounted in the Rear-View Mirror

This four-element mirror array is mounted in the rear-view mirror. AKG directional microphones of type cardioid are used. This arrangement is known to be compatible with product design constraints, since it belongs to the standard speech input equipment in the Mercedes E class. The distance from the driver mouth to the mirror depends on the position of the driver seat. This distance can amount between 50 and 80 cm, a typical mouth–microphone distance being about 60 cm. The experimental setup with the driver and the codriver is depicted in Fig. A.2.

A.2 The Two-Element Distributed Array Mounted on the Car Ceiling

The two-element distributed array consists of directional microphones oriented to the driver and the codriver, respectively. The two microphones are placed on roof control panel with 17-cm spacing. In contrast to the four-element compact array mounted in the rear-view mirror, this microphone arrangement provides the interferer reference signal directly. The experimental setup is depicted in Fig. A.3. For this setup, we use PEIKER directional microphones of type cardioid.

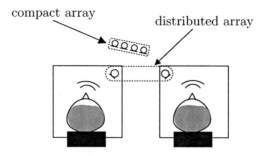

Fig. A.1. Experimental layouts for the car

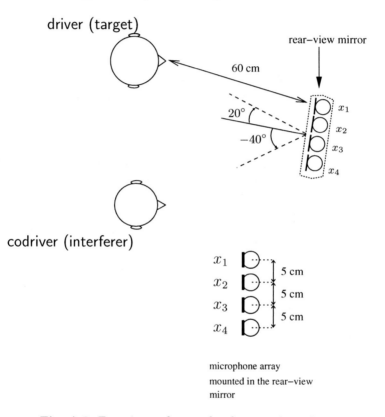

Fig. A.2. Experimental setup for the rear-view mirror

Note that it is straightforward to extend this two-element array to a four-element array by placing directional microphone on the roof, close to each of the two back-seat passengers. This is of particular interest when separating the driver speech from the interfering back-seat left passenger, since the four-element compact array mounted in the rear-view mirror may fail in this case.

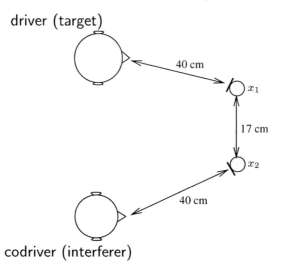

driver (target)

40 cm

x_1

17 cm

x_2

40 cm

codriver (interferer)

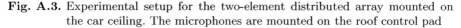

Fig. A.3. Experimental setup for the two-element distributed array mounted on the car ceiling. The microphones are mounted on the roof control pad

A.3 Acoustic Characteristics of the Car Cabin

Room impulse response

The car cabin is usually regarded as weakly reverberant. The reverberation time, which is denoted by T_{60} and is defined as the time which is necessary for the sound energy to decrease by 60 dB [40], is about $T_{60} = 50$ ms.

Estimating the impulse response between the driver mouth and a given microphone can be done using an artificial head. To avoid the influence of the loudspeaker, we use a close-talk microphone mounted on the artificial head. Figure A.4 shows the impulse response between this close-talk microphone and the microphone x_1 of the four-element compact array mounted in the rear-view mirror.

Background noise

Car noise consists of motor noise and also results from the wind and from the contact between the tires and the road. It can be regarded as diffuse and has its main energy in low frequency bands [14]. The PSD of road noise that was recorded at 100 km h^{-1} is shown in Fig. A.5.

A.4 Illustration of the Difficulty in the Design of a Reliable DTD

An experiment is conducted in the car interior with two male speakers recorded with a four-element microphone array mounted in the rear-view

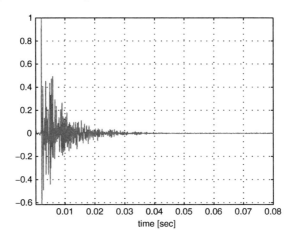

Fig. A.4. Impulse response estimated between a close-talk microphone mounted on the artificial head on the driver seat and the microphone x_1 of the four-element compact array mounted in the rear-view mirror

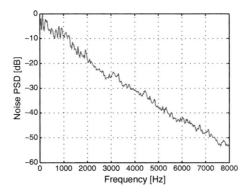

Fig. A.5. PSD of a typical road noise recorded at 100 km h^{-1}. The PSD is estimated using the Welch periodogram averaged over 160,000 samples (FFT length $NFFT = 1024$, sample frequency $f_s = 16,000$ Hz)

mirror.[1] The speakers are positioned on the driver and codriver seats, respectively. We observe the power of the source signals, which indicates at what time each speaker is active. On the other hand, we examine the input SIR estimation

$$\mathrm{SIR}_{\mathrm{est}}(p) = \frac{(M-1)x_0^2(p)}{\sum_{m'=0}^{M-1} x_{B,m'}^2(p)} \qquad (A.1)$$

---

[1] We refer to Appendix A for a detailed description of the experimental setup, which is depicted in Figs. A.1 and A.2.

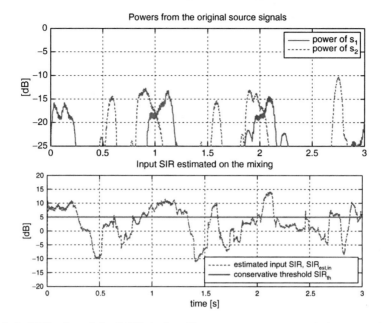

Fig. A.6. Difficulty of the DTD design. Two sources were recorded separately. The *upper plot* shows the individual signal powers. The *lower plot* shows the estimated input SIR, $\mathrm{SIR_{est,in}}(p)$. The represented quantities are averaged over a sliding window of length 30 ms

We are interested in finding a threshold $\mathrm{SIR_{th}}$ such that (1) $\mathrm{SIR_{th}} < \mathrm{SIR_{est,in}}(p)$ during target silences and interferer activity and (2) $\mathrm{SIR_{th}} > \mathrm{SIR_{est,in}}(p)$ during target activity. As can be see from Fig. A.6, such a threshold does not exist in general: The estimate $\mathrm{SIR_{est,in}}(p)$ can reach similar levels during target and interferer activity. Hence, a trade-off must be found between

- robustness: choosing a low, conservative threshold that prevents target cancelation and
- accuracy in the detection of interferer activity: choosing a higher threshold that allows adaptation more often.

Obviously, this issue worsens with increasing overlap of the target and interferer speech. This simple example has illustrated the difficulties that arise with the design of a reliable DTD.

B

Far- and Free-Field Acoustic Propagation Model and Null Beamforming

B.1 Far- and Free-Field Model

We consider a source at position (r, ϕ, θ) in spherical coordinates. Without loss of generality, the reference is taken at the position of the microphone $\mathbf{x}_1$. Next, we assume a uniform linear array (ULA) with intermicrophone spacing Δ arranged along the z-axis as shown in Fig. B.1. Then the position of the source relative to the array is invariant by rotational symmetry of angle θ. Let us denote the distance from the source $\mathbf{s}$ to the second microphone $\mathbf{x}_2$ by r_2. We have

$$r_2^2 = r^2 + \Delta^2 + 2\Delta r \sin \theta. \tag{B.1}$$

We next compute the length $r_2 - r$ of the path traveled by the sound wave when propagating from the reference microphone to the next one. Using (B.1), we find that this additional path is given by

$$r_2 - r = r \left(\sqrt{1 + \left(\frac{\Delta}{r}\right)^2 + 2\frac{\Delta}{r} \sin \theta} - 1 \right). \tag{B.2}$$

The far-field model corresponds to the situation $r \gg \Delta$. Using the first-order development $\sqrt{1 + x} \approx 1 + \frac{x}{2} + o(x^2)$, we can compute the limit of (B.2) for $r \to +\infty$:

$$\lim_{r \to +\infty} (r_2 - r) = \Delta \sin \theta. \tag{B.3}$$

Therefore the time τ_θ needed by the sound wave to travel from one sensor to the next one if given for the far- and free-field model by

$$\tau_\theta = \frac{\Delta \sin \theta}{c}, \tag{B.4}$$

where c denotes the speed of sound.

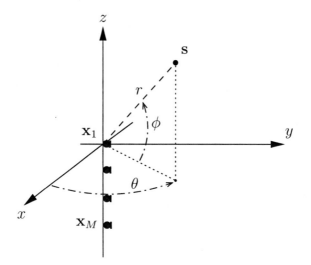

Fig. B.1. Cartesian and spherical coordinate systems

B.2 Null Beamforming

In the following we explain how to compute filter coefficients $\mathbf{w}$ with a unit response in the direction θ_1 and a zero response in the direction θ_2. Without loss of generality, we assume that $\theta_1 = 0$, which is equivalent to steering the array toward θ_1.

We want to find the MISO separation system $\mathbf{w}$ that fulfills

$$\begin{cases} \mathbf{g}(\mathbf{w}, 0) = \delta_D, \\ \mathbf{g}(\mathbf{w}, \theta_2) = \mathbf{0}, \end{cases} \tag{B.5}$$

where $\mathbf{g}(\mathbf{w}, \theta)$ denotes the spatial response of $\mathbf{w}$ in the direction θ, as defined in (2.27). To simplify the presentation, we assume that the delay $\tau_{\theta_2} = f_s \frac{\Delta}{c} \sin \theta_2$ is an integer. Then the pair of equations in (B.5) may be reformulated as follows:

$$\begin{cases} \sum_{m=1}^{M} w_{m,k} = \delta_{D,k} & \text{for } k = 1, \dots, L, \\ \sum_{m=1}^{M} w_{m,k+(m-1)\tau_{\theta_2}} = 0 & \text{for } k = 1, \dots, L, \end{cases} \tag{B.6}$$

where $\delta_{D,k} = 1$ for $k = D$ and $\delta_{D,k} = 0$ otherwise. The lower equation in (B.6), $\sum_m w_{m,k+(m-1)\tau_{\theta_2}} = 0$, characterizes null beamformers which have a zero spatial response at θ_2. The spatial response of our null beamformer is also specified in the target direction $\theta_1 = 0$. To obtain a matrix form of (B.6), we introduce the matrix $\mathbf{R_{nn}}$ as

$$\mathbf{R_{nn}} \triangleq \begin{bmatrix} \mathbf{I}_{L \times L} & \mathbf{D}_{\tau_{\theta_2}} & \cdots & \mathbf{D}_{(M-1)\tau_{\theta_2}} \\ \mathbf{D}_{-\tau_{\theta_2}} & \mathbf{I}_{L \times L} & \cdots & \mathbf{D}_{(M-2)\tau_{\theta_2}} \\ \vdots & & \ddots & \vdots \\ \mathbf{D}_{-(M-1)\tau_{\theta_2}} & \cdots & \cdots & \mathbf{I}_{L \times L} \end{bmatrix}. \tag{B.7}$$

$\mathbf{D}_\tau$ is a square matrix containing zeros, except for the τth upper diagonal that contains ones. In fact $\mathbf{R_{nn}}$ corresponds to the microphone correlation matrix $\mathbf{E}\left\{\mathbf{x}(p)\mathbf{x}^T(p)\right\}$ when the source at θ_2 emits a stationary white noise. Equation (B.6) may now be written as

$$\mathbf{w}^T \left[\mathbf{C}\,\mathbf{R_{nn}}\right] = \left[\boldsymbol{\delta}_0^T\; \mathbf{0}_{1 \times ML}\right] \tag{B.8}$$

in matrix form. (The matrix $\mathbf{C}$ was given in (3.24).) If there are more than two microphones, the $2L$ equations (B.6) form an under determined set of constraints for the ML filter coefficients. If we simultaneously constrain $\mathbf{w}$ to minimize the white-noise gain, the filters $\mathbf{w}$ are given by the pseudoinverse

$$\mathbf{w}^T = \left[\boldsymbol{\delta}_0^T\; \mathbf{0}_{1 \times ML}\right]\left[\mathbf{C}\,\mathbf{R_{nn}}\right]^+. \tag{B.9}$$

In general, $[\mathbf{C}\,\mathbf{R_{nn}}]$ has neither full column rank nor full row rank. The computation of the pseudoinverse involves its singular value decomposition and it is difficult to find a general closed formula for $\mathbf{w}$. For this reason, the solution of (B.9) may rather be found numerically.

C

The RGSC According to Hoshuyama et al.

This appendix describes the robust GSC (RGSC) proposed by Hoshuyama et al. [53]. Using an adaptation control mechanism, the RGSC can be an efficient LCMV beamformer because it adapts the constraint of distortionless transmission of the desired source to the current acoustic environment [47]. The RGSC includes an adaptive blocking matrix to minimize the leakage of the target signal. The adaptive blocking matrix consists of a set of "interference cancelers" implemented as adaptive filters. As shown by Herbort et al. [47], the RGSC is an LCMV beamformer with a relaxed constraint and may yield an improved suppression of the interference signal relative to the original GSC with a fixed blocking matrix [42]. Section C.1 describes the RGSC which can be used in conjunction with the four-element compact array mounted in the rear-view mirror. In Sect. C.2, we modify this RGSC for use with the two-element distributed array mounted on the car ceiling.

C.1 RGSC for the Four-Element Compact Array Mounted in the Rear-View Mirror

The blocking matrix is square with M outputs instead of $M-1$ in the original GSC, and has a particular form. To obtain an expression of its outputs, we need to define the filters $b_m, m = 1, \ldots, M$ of length[1] L. The coefficients of the filter b_m are stacked in the $L \times 1$ vector:

$$\mathbf{b}_m \triangleq (b_{m,0}, \ldots, b_{m,L-1})^{\mathrm{T}}. \tag{C.1}$$

---

[1] The filter length of b_m is not necessarily equal to the filter length of the interference canceler. Let us temporarily denote the length of b_m by L_B. Maximum robustness against leakage is obtained for $L_B \geq L$, since if $L_B \geq L$ then the interference canceler is not long enough to access any target component. To minimize the complexity, we set $L_B = L$.

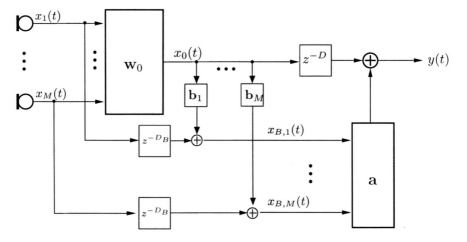

Fig. C.1. Robust generalized sidelobe canceler (RGSC) according to Hoshuyama et al.

We also need to define the vector $\mathbf{x}_0(p)$ for the target reference signal:

$$\mathbf{x}_0(p) \triangleq (x_0(p), \ldots, x_0(p - L + 1))^{\mathrm{T}}. \qquad (C.2)$$

The outputs of the blocking matrix are then given by

$$x_{B,m}(p) = x_m(p - D_B) + \mathbf{b}_m^{\mathrm{T}} \mathbf{x}_0(p), \qquad (C.3)$$

as depicted in Fig. C.1. As shown in (C.8), the optimum filters $\mathbf{b}_m$ have to model the inverse of a sum of room impulse responses. Since the room impulse responses may be nonminimum-phased, modeling it inverse requires an acausal delay [50]. Therefore, the delay D_B is typically set to $D_B = L/2$. It must also be included in the interference canceler, which yields $D = L/2 + D_B = L$.

The filters $\mathbf{b}_m$ are adapted so that the target leakage at the output of the blocking matrix is minimum. In practice, the adaptation is carried out when the desired source is dominant in such a way that the variance $\mathbf{E}\left\{x_{B,m}^2(p)\right\}$ is minimized. This is achieved with the Wiener solution:

$$\mathbf{b}_m = -\left(\mathbf{E}\left\{\mathbf{x}_0(p)\mathbf{x}_0^{\mathrm{T}}(p)\right\}\right)^{-1} \mathbf{E}\left\{x_m(p - D_B)\mathbf{x}_0(p)\right\}. \qquad (C.4)$$

In an adaptive context, the filters $\mathbf{b}_m$ can be adapted with the NLMS algorithm as follows:

$$\mathbf{b}_m(p + 1) = \mathbf{b}_m(p) - \mu_{B,NLMS} \frac{x_{B,m}(p)\mathbf{x}_0(p)}{\|\mathbf{x}_0(p)\|^2}. \qquad (C.5)$$

For our experiments with speech signals, the step-size has been set to $\mu_{B,NLMS} = 0.1$. The delay in the blocking matrix path is $D_B = L/2$. The delay in the fixed beamformer path is $D = D_B + L/2 = L$.

Representation in the DTFT domain

Let us denote by $H_m(\omega)$ the transfer function from the desired source $S(\omega)$ to the mth microphone $X_m(\omega)$ and set $\mathbf{H}(\omega) = (H_1(\omega), \ldots, H_M(\omega))^T$. In the noiseless case, the source–microphone relationship is written in the DTFT domain as:

$$\mathbf{X}(\omega) = \mathbf{H}(\omega)S(\omega). \tag{C.6}$$

We denote the DTFT of the filter b_m by $B_m(\omega)$ and we define the vector $\mathbf{B}(\omega) \triangleq (B_1(\omega), \ldots, B_M(\omega))^T$. The filters $B_m(\omega)$ are adapted so that the target leakage at the blocking matrix output $X_{B,m}(\omega) = e^{-i\omega D_B}X_m(\omega) + B_m(\omega)X_0(\omega)$ is minimum. It can be shown that this leads to a *signal-dependent* spatial constraint $\mathbf{W}^H(\omega)\mathbf{H}(\omega)S(\omega) = S(\omega)$ [47]. If the desired signal has no energy in a given frequency band, the constraint vanishes, yielding $\mathbf{B}(\omega) = \mathbf{I}$. This allows the filters $\mathbf{W}(\omega) = 0$ in that band. This would not be possible if the spatial constraint was maintained across the entire spectrum. The filters $B_m(\omega)$ that minimize $\mathbf{E}\left\{|X_{B,m}(\omega)|^2\right\}$ are obtained by the Wiener solution:

$$B_m(\omega) = -e^{-i\omega D_B}\Phi_{X_m X_0}(\omega)/\Phi_{X_0}(\omega). \tag{C.7}$$

$\Phi_{X_m X_0}$ denotes the CPSD of the signals x_m and x_0. Likewise, Φ_{X_0} denotes the PSD of the signal x_0. Combining (C.7) and (C.6) for $S_m(\omega) \neq 0$ yields the optimal solution in the minimum mean square error (MMSE) sense

$$B_m(\omega) = -e^{-i\omega D_B}H_m(\omega)/\mathbf{W}_0(\omega)\mathbf{H}(\omega). \tag{C.8}$$

Equation (C.8) shows that the optimal blocking matrix filters $B_m(\omega)$ involve the inverse of the sum of delayed acoustic transfer functions, since $\mathbf{W}_0^H(\omega)\mathbf{H}(\omega) = \sum_{m=1}^{M} e^{i\omega(m-1)\tau_\theta}H_m(\omega)$.

C.2 RGSC for the Two-Element Distributed Array Mounted on the Car Ceiling

A data-dependent "blocking matrix" can be considered for the two-element distributed array mounted on the car ceiling, too. Using a set of interference cancelers for the blocking matrix as in Sect. C.1, we can incorporate an additional adaptive filter in the GSC structure. This transforms the AIC shown in Fig. C.2 a into the RGSC shown in Fig. C.2 b. The outputs $x_{B,m}(p)$ of the blocking matrix are defined as

$$x_{B,m}(p) \triangleq x_{m+1}(p) - \mathbf{b}_m^T \mathbf{x}_1(p) \quad \text{for } m = 1, \ldots, M-1. \tag{C.9}$$

The filters $\mathbf{b}_m$ should be adapted using interference-free input signal. They can be computed with (C.4) or (C.5) using $\mathbf{x}_0(p) = \mathbf{x}_1(p)$ and $\mathbf{x}_{B,m}(p) = \mathbf{x}_{m+1}(p)$.

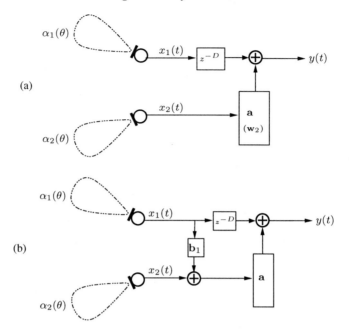

Fig. C.2. Beamformer architecture for two directional microphones. (a) Adaptive interference canceler (AIC). (b) RGSC according to [53] modified for use with the two-element distributed array mounted on the car ceiling

C.3 Experimental Comparison: GSC vs. RGSC

Offline experiments in stationary conditions

This section examines how the RGSC mitigates the target-cancelation problem. In the following, the performance of the RGSC is compared with that of the GSC in stationary conditions with nonadaptive, off-line beamformers. That way, the comparison should remain independent of the implementation. By contrast, the beamformer performance in adaptive mode depends highly on implementation parameters like the step-size and the DTD thresholds.

The source signals are white-noise signals emitted by artificial heads to ensure stationary conditions. The microphone signals are sampled at $f_s = 16\,\mathrm{kHz}$. The filter length is set to $L = 256$ in the case of the four-element compact array mounted in the rear-view mirror and to $L = 512$ in the case of the distributed array. The experimental setups are described in Appendix A in greater detail.

Varying parameters

To evidence how the adaptive blocking matrix mitigates the power-inversion effect, we let vary the SIR of the input signals that are used to adapt the interference canceler, which is denoted by SIR_{in}. To define SIR_{in}, we decompose the input signals as the sum of the contribution of the desired source and of that of the interferers, as given in Sect. 2.4:

$$x_m(p) = x_{\text{sig},m}(p) + x_{\text{int},m}(p). \tag{2.61}$$

In offline mode with stationary signals, we use a batch estimate of the expectation operator:

$$\widehat{\mathbf{E}}\{f(\mathbf{A}(p))\} \triangleq \frac{1}{T}\sum_{p=1}^{T} f(\mathbf{A}(p)). \tag{2.65}$$

Using (2.61), we may define SIR_{in} and SIR_{in}^d for compact and distributed arrays respectively, as follows:

$$\text{SIR}_{\text{in}} \triangleq \frac{\sum_{m=1}^{M} \widehat{\mathbf{E}}\{x_{\text{sig},m}^2(p)\}}{\sum_{m=1}^{M} \widehat{\mathbf{E}}\{x_{\text{int},m}^2(p)\}}, \qquad \text{SIR}_{\text{in}}^d \triangleq \frac{\widehat{\mathbf{E}}\{x_{\text{sig},1}^2(p)\}}{\widehat{\mathbf{E}}\{x_{\text{sig},1}^2(p)\}}. \tag{C.10}$$

For a given blocking matrix $\mathbf{B}$, we also decompose the interference reference signals as the sum of the contribution of the desired source and of that of the interferers:

$$\mathbf{x}_{B,\text{sig},m}(p) \triangleq \mathbf{B}^{\text{T}}\mathbf{x}_{\text{sig},m}(p), \qquad \mathbf{x}_{B,\text{int},m}(p) \triangleq \mathbf{B}^{\text{T}}\mathbf{x}_{\text{int},m}(p). \tag{C.11}$$

Using (C.11), the SIR at the input of the interference canceler, $\text{SIR}_{\text{AIC,in}}$, (that is, the SIR at the output of the blocking matrix) may defined as

$$\text{SIR}_{\text{AIC,in}} \triangleq \frac{\sum_{m=1}^{M} \widehat{\mathbf{E}}\{x_{B,\text{sig},m}^2(p)\}}{\sum_{m=1}^{M} \widehat{\mathbf{E}}\{x_{B,\text{int},m}^2(p)\}}, \qquad \text{SIR}_{\text{AIC,in}} \triangleq \frac{\widehat{\mathbf{E}}\{x_{B,\text{sig},1}^2(p)\}}{\widehat{\mathbf{E}}\{x_{B,\text{int},1}^2(p)\}}. \tag{C.12}$$

$\text{SIR}_{\text{AIC,in}}$ may be obtained by subtracting a fixed offset a from SIR_{in} in the dB scale:

$$\text{SIR}_{\text{AIC,in}} = \text{SIR}_{\text{in}} - a. \tag{C.13}$$

The offset a depends on the blocking matrix $\mathbf{B}$. According to the power-inversion effect described in Sect. 3.4, the SIR at the output of the beamformer should be zero for $\text{SIR}_{\text{in}} = a$.

Filter adaptation

The filter adaptation and the performance evaluation are carried out in three steps:

(i) First the RGSC filters $\mathbf{b}_m$ for $m = 1,\dots,M$ of the data-dependent blocking matrix are computed according to (C.4). These filters $\mathbf{b}_m$ are trained using the desired source input signals $\mathbf{d}(p)$, which corresponds to a best-case scenario. For the GSC, the blocking matrix is fixed.

(*ii*) Then the interference canceler **a** is adapted on the input signals having a prescribed SIR_{in} (SIR at the input of the beamformer). The adaptation of the interference canceler **a** is based on the Wiener solution (3.22). However, the inversion of $\mathbf{R}_{\mathbf{x}_B\mathbf{x}_B}$ in (3.22) for $L = 256$ is numerically badly conditioned. Therefore, we regularize the inversion with a regularization factor $\epsilon > 0$, as follows:

$$\mathbf{a}^{(0)} = (\mathbf{R}_{\mathbf{x}_B\mathbf{x}_B} + \epsilon\mathbf{I})^{-1}\,\mathbf{E}\left\{x_0(p - D)\mathbf{x}_B(p)\right\}. \tag{C.14}$$

This yields a biased first estimate $\mathbf{a}^{(0)}$ of the optimal **a**. For our experiments, we set $\epsilon = 0.01$ which led to the best results. Then, we improve $\mathbf{a}^{(0)}$ using the gradient descent for the cost function in (3.1). The gradient descent does not necessitate matrix inversion. Using the gradient in (3.21) yields

$$\mathbf{a}^{(n+1)} = \mathbf{a}^{(n)} - \mu\frac{\partial J}{\partial \mathbf{a}} \tag{C.15}$$

$$= \mathbf{a}^{(n)} - 2\mu\sum_{p=1}^{T} y(p)\mathbf{x}_B(p). \tag{C.16}$$

The step-size μ is set to $\mu = 0.05$ and the iterations are carried out until J does not decrease by more than $0.1\,\text{dB}$ per iteration.

(*iii*) After (*i*) and (*ii*), the filters are fixed and the performances of the full GSC and RGSC structures are evaluated using unit power input signals from the target and from the interferer.

C.3.1 Experiments with the Four-Element Compact Array Mounted in the Rear-View Mirror

We consider first the four-element directional microphone array mounted in the rear-view mirror (see Sect. A.1) for more details). The delay D is set to $D = L/2$. The RGSC filters $\mathbf{b}_m$ for $m = 1, \ldots, M$ of the data-dependent blocking matrix are trained using the desired source input signals $d(p)$. It could be measured that this data-dependent blocking matrix brings a $11.5\,\text{dB}$ reduction of the SIR, that is, $a = 11.5\,\text{dB}$ in (C.13).

For our implementation of the GSC, the output signals of the fixed blocking matrix $x_{B,m}(p)$ are given by

$$x_{B,m}(p) = x_{m+1}(p) - x_0(p) \quad \text{for } m = 1, \ldots, M-1, \tag{C.17}$$

$$= x_{m+1}(p) - \frac{1}{M}\sum_{m=1}^{M-1} x_{M-1}(p). \tag{C.18}$$

We could measure that the signal $x_{B,m}(p)$ exhibits a SIR which is about $3\,\text{dB}$ lower that the input SIR, that is, $a = 3\,\text{dB}$ in (C.13). The results are shown in Fig. C.3.

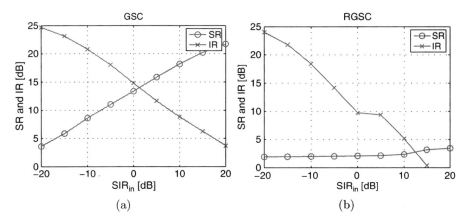

Fig. C.3. Performances of the GSC (**a**) and the RGSC (**b**) as a function of SIR_{in} (SIR at the beamformer input), for the four-element compact array mounted in the rear-view mirror. The RGSC beamformer structure is shown in Fig. C.1. For the RGSC, the blocking matrix is trained with interferer-free input signals, which is the best-case scenario. The experimental setup is depicted in Fig. A.2. The steered DOA is $\theta = 20°$. The filter length is set to $L = 256$

It can be observed that the GSC does not bring any SIR improvement for SIR_{in} around 3 dB. This observation fits well with the power-inversion effect. Moreover we can see that the level of the desired signal is significantly reduced for large SIR_{in}. On the other hand, the data-dependent blocking-matrix attenuates the desired source components at the outputs $x_{B,m}$ of the blocking matrix. This protects efficiently against cancelation of the target signal: We can see that the reduction of the desired signal level is much smaller than that of the GSC. However, in accordance with the power-inversion effect, the SIR improvement vanishes for $SIR_{in} = 11.5$ dB: The signals $x_{B,m}$ still contain reflected paths of the desired source and are correlated to the output of the fixed beamformer x_0. Since the optimal interference canceler minimizes this correlation, an important degradation of the interference reduction results for $SIR_{in} > -5$ dB.

C.3.2 Experiments with the Two-Element Distributed Array Mounted on the Car Ceiling

We compare the AIC structure and the RGSC structure with a data-dependent blocking matrix. Both structures are depicted in Fig. 3.2. We could measure that the SIR at microphone x_2 is about 5 dB below the SIR at microphone x_1, that is, $a = 5$ dB in (C.13). The blocking matrix filter $\mathbf{b}_1$ is adapted with the desired source signals $\mathbf{d}(p)$, which represents a best-case scenario. This data-dependent blocking matrix brings an additional 7 dB reduction of

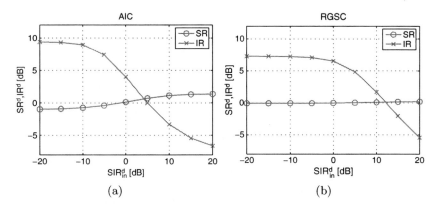

Fig. C.4. Performances of the AIC (**a**) and of the RGSC (**b**) as a function of SIR_{in}^c (SIR at the beamformer input), for the two-element distributed array mounted on the car ceiling. The two beamformer structures are shown in Fig. C.2. In the case of the RGSC, the blocking matrix filter $\mathbf{b}_1$ is trained using interferer-free input signals, which is the best-case scenario. The experimental setup is depicted in Fig. A.3. The filter length is set to $L = 512$

the SIR with respect to the SIR at microphone x_2, that is, $a = 12\,\mathrm{dB}$ in (C.13). Figure C.4 shows the SIR improvement for various input SIR. The RGSC is slightly more robust against signal cancelation, but both structures exhibit similar SR^c curves. Although these results do not account for the distortion of the target signal, they indicate that the target signal cancelation problem is not significantly better relieved by the RGSC than by the AIC.

This may be explained by the causality constraints than can be set on the AIC (that is, the AIC structure with $D = 0$). For the free-field propagation model, the target signal with positive DOA reaches the microphone $x_2(p)$ after $x_1(p)$, i.e., after a positive delay. Then, causal filtering $\mathbf{a}^T\mathbf{x}_2(p)$ cannot compensate this delay to suppress the target at the output $y(p) = x_1(p) + \mathbf{a}^T\mathbf{x}_2(p)$. This may prevent the desired source to be canceled. As a consequence, the AIC is chosen for the two-element distributed array mounted on the car ceiling.

C.4 Conclusion

The RGSC proposed by Hoshuyama et al. [53] has been described. A modification of this RGSC for use in conjunction with the two-element distributed array mounted on the car ceiling has been proposed.

In comparison to the original GSC [42] for the four-element compact array mounted in the rear-view mirror, the M outputs of the adaptive blocking matrix may provide more degrees of freedom to the interference canceler, depending on the target signal spectrum. This may yield an improved suppression of the interferer signal at the beamformer output. Moreover, the RGSC provides a significant improvement over the GSC in terms of robustness against target signal cancelation. Therefore, in our experiments, the RGSC is chosen to determine the performance of controlled beamforming.

For the two-element distributed array mounted on the car ceiling, the advantage of the RGSC over the AIC is less significant than for the four-element compact array mounted in the rear-view mirror. The increased computational demand of the RGSC over the AIC may not be justified. For this reason, we prefer to use the AIC to determine the performance of controlled beamforming in our experiments.

D

Stability Analysis

As explained in Sect. 7.1.1, the analysis of the *global* stability for convolutive BSS is hardly tractable. This chapter examines the *local* stability in the vicinity of a particular equilibrium point. The local stability of BSS algorithms has been investigated in the vicinity of the inverse of the mixing system [8, 26, 39, 52]. There, the sources are not only separated but also deconvolved, possibly using a feedback filter architecture, as in [26]. However, in many practical cases, only the separation of the sources is desired (or achievable). Therefore, we will study the stability around an equilibrium point that separates but does not deconvolve the sources.

This chapter constitutes a generalization of the analysis for white source signals presented in [16]. It is organized as follows: Section D.1 defines the mixing and separation models and the notations. Section D.2 derives the linearization of the learning rules in the vicinity of the equilibrium. The conditions for the local stability are derived in Sect. D.3.

D.1 Mixing and Separation Models

We consider a 2×2 mixing and separation scenario with mixing and separation filters of the same lengths ($L_{\mathrm{m}} = L$). We assume that the diagonal channels of the mixing and separation systems are delayed unit responses, that is,

$$\mathbf{h}_{11} = \mathbf{h}_{22} = \boldsymbol{\delta}_d, \tag{D.1}$$

for some delay $0 \le d \le L/2$. Note that this assumption is only a normalization of the mixing process. The off-diagonal channels are denoted by $\mathbf{h}_{12}$ and $\mathbf{h}_{21}$. We constrain accordingly the diagonal filters of the separation system

$$\mathbf{w}_{11} = \mathbf{w}_{22} = \boldsymbol{\delta}_d. \tag{D.2}$$

This model is depicted in Fig. 2.6. It approximates a scenario where each source $s_n(p)$ is positioned close to microphone $x_n(p)$. It is also motivated by the following remarks:

(i) Let us consider a small deviation ε around the equilibrium point $\mathbf{W}_{\text{opt}}$, that is, $\mathbf{W} = \mathbf{W}_{\text{opt}} + \varepsilon$. We parametrize the deviation as $\varepsilon^C(n) = \varepsilon(n)\mathbf{H}$. $\varepsilon^C(n)$ corresponds to an additive deviation for the global system $\mathbf{C}(n) = \mathbf{W}(n)\mathbf{H}$, since $\mathbf{C}(n) = \mathbf{W}_{\text{opt}}\mathbf{H} + \varepsilon^C(n)$. It can be shown that the diagonal components $\varepsilon_{ii}^C(n)$ of $\varepsilon^C(n)$ are constant at first order [33]. Therefore, the diagonal elements of the global system $\mathbf{C}_{ii}(n)$ are not going to change significantly and $\mathbf{W}(n)$ will not reach the vicinity of $\mathbf{W}_{\text{opt}}$. The constraint $w_{ii} = \delta_d$ for $i = 1, \ldots, N$ provides additional information that enable convergence toward the vicinity of $\mathbf{W}_{\text{opt}}$.

(ii) The constraints $\mathbf{w}_{ii} = \delta_d$ for $i = 1, 2$ ensure the uniqueness of the equilibrium point, which is $\mathbf{w}_{ij} = -\mathbf{h}_{ij}$ for $i, j = 1, 2$ and $i \neq j$.

(iii) Under the constraint $\mathbf{w}_{ii} = \delta_d$ for $i = 1, 2$, the self-closed learning rules (6.51) are equal to their self-closed counterparts (6.53).

This mixing system is separated with the following separation system:

$$\begin{aligned} \mathbf{w}_{11} &= \delta_d, \quad \mathbf{w}_{12} = -\mathbf{h}_{12}, \\ \mathbf{w}_{21} &= -\mathbf{h}_{21}, \; \mathbf{w}_{22} = \delta_d. \end{aligned} \tag{D.3}$$

For later reference, we define the vector $\mathbf{w}_{\text{eq}}$ of length $2L - 1$ as

$$\mathbf{w}_{\text{eq}} = \delta_{2d} - \mathbf{h}_{12} * \mathbf{h}_{21}. \tag{D.4}$$

In fact, $\mathbf{w}_{\text{eq}} = (w_{\text{eq},0}, \ldots, w_{\text{eq},2L-2})^{\mathrm{T}}$ is the source–output response at the solution (D.3). The sources s_1 and s_2 are assumed stationary within time blocks of length L and their self-correlation function is denoted by

$$r_{n,\tau}(p) = \mathbf{E}\left\{s_n(p)s_n(p - \tau)\right\}. \tag{D.5}$$

D.2 Linearization of the NG-SOS-BSS Updates

We examine the local stability of the separation algorithm around the separating solution (D.3). To this end, we set

$$\mathbf{w}_{12}(n) = -\mathbf{h}_{12} + \varepsilon_{12}(n), \tag{D.6}$$

$$\mathbf{w}_{21}(n) = -\mathbf{h}_{21} + \varepsilon_{21}(n), \tag{D.7}$$

with the deviations

$$\varepsilon_{12} = (\varepsilon_{12,0}, \ldots, \varepsilon_{12,L-1})^{\mathrm{T}}, \tag{D.8}$$

$$\varepsilon_{21} = (\varepsilon_{21,0}, \ldots, \varepsilon_{21,L-1})^{\mathrm{T}}. \tag{D.9}$$

The expectation of algorithm (6.51) can be written as:

$$\mathbf{w}_{12}(n + 1) = \mathbf{w}_{12}(n) - \mu \sum_{k=1}^{K} \mathbf{r}_{\mathbf{y}_1\mathbf{y}_2}(kL)/\sigma_1^2(kL), \tag{D.10}$$

$$\mathbf{w}_{21}(n + 1) = \mathbf{w}_{21}(n) - \mu \sum_{k=1}^{K} \mathbf{r}_{\mathbf{y}_2\mathbf{y}_1}(kL)/\sigma_2^2(kL), \tag{D.11}$$

where

$$\mathbf{r}_{\mathbf{y}_i\mathbf{y}_j}(p) = \left(r_{\mathbf{y}_i\mathbf{y}_j,-d}(p), \ldots, r_{\mathbf{y}_i\mathbf{y}_j,L-1-d}(p)\right) \tag{D.12}$$

$$r_{\mathbf{y}_i\mathbf{y}_j,\tau}(p) = \mathbf{E}\left\{y_i(p)y_j(p-\tau)\right\} \tag{D.13}$$

$$\sigma_i^2(p) = \mathbf{E}\left\{y_i^2(p)\right\} \tag{D.14}$$

for $i,j = 1,2$. The BSS algorithm is driven by the output cross correlation $\mathbf{r}_{\mathbf{y}_i\mathbf{y}_j}(p)$. Neglecting the quadratic terms in $\varepsilon_{ij,k}(n)$, this cross correlation is first-order approximated by

$$r_{\mathbf{y}_1\mathbf{y}_2,\tau}(p) = \sum_{u=0}^{2L-2}\sum_{v=0}^{L-1} w_{\text{eq},u}\varepsilon_{21,v}(n)r_{1,v+d+\tau-u}(p)$$

$$+ \sum_{u=0}^{L-1}\sum_{v=0}^{2L-2} w_{\text{eq},v}\varepsilon_{12,u}(n)r_{2,v+\tau-u-d}(p). \tag{D.15}$$

The first-order approximation for $r_{\mathbf{y}_2\mathbf{y}_1,\tau}(p)$ is simply given by

$$r_{\mathbf{y}_2\mathbf{y}_1,\tau}(p) = r_{\mathbf{y}_1\mathbf{y}_2,-\tau}(p). \tag{D.16}$$

Substituting $\mathbf{w}_{ij}(n)$ in (D.6) and (D.7) into (D.10) and (D.11), and replacing $r_{\mathbf{y}_i\mathbf{y}_j,\tau}(p)$ by its value in (D.15), the deviations $\varepsilon_{12}(n+1)$ and $\varepsilon_{21}(n+1)$ are given as a linear function of $\varepsilon_{12}(n)$ and $\varepsilon_{21}(n)$. Thus, the first-order approximation of (D.10) and (D.11) can be written in terms of $\varepsilon_{12}(n)$ and $\varepsilon_{21}(n)$ in matrix form:

$$\varepsilon_{12}(n+1) = \varepsilon_{12}(n) - \mu\sum_k\left(\frac{1}{\sigma_1^2(kL)}\mathbf{A}_{11}(kL)\varepsilon_{12}(n) + \frac{1}{\sigma_1^2(kL)}\mathbf{A}_{12}(kL)\varepsilon_{21}(n)\right),$$
$$\tag{D.17}$$

$$\varepsilon_{21}(n+1) = \varepsilon_{21}(n) - \mu\sum_k\left(\frac{1}{\sigma_2^2(kL)}\mathbf{A}_{21}(kL)\varepsilon_{12}(n) + \frac{1}{\sigma_2^2(kL)}\mathbf{A}_{22}(kL)\varepsilon_{21}(n)\right)$$
$$\tag{D.18}$$

with

$$\mathbf{A}_{11}(p) = \begin{pmatrix} \tilde{w}_{\text{eq},0}^{(2)}(p) & \tilde{w}_{\text{eq},1}^{(2)}(p) & \ldots & \tilde{w}_{\text{eq},L-1}^{(2)}(p) \\ \tilde{w}_{\text{eq},-1}^{(2)}(p) & \tilde{w}_{\text{eq},0}^{(2)}(p) & \ldots & \tilde{w}_{\text{eq},L-2}^{(2)}(p) \\ \vdots & & \ddots & \vdots \\ \tilde{w}_{\text{eq},-L+1}^{(2)}(p) & \ldots & \ldots & \tilde{w}_{\text{eq},0}^{(2)}(p) \end{pmatrix}, \tag{D.19}$$

$$\mathbf{A}_{12}(p) = \begin{pmatrix} w^{(1)}_{\mathrm{eq},0}(p) & w^{(1)}_{\mathrm{eq},1}(p) & \cdots & w^{(1)}_{\mathrm{eq},L-1}(p) \\ w^{(1)}_{\mathrm{eq},1}(p) & w^{(1)}_{\mathrm{eq},2}(p) & \cdots & w^{(1)}_{\mathrm{eq},L}(p) \\ \vdots & & & \vdots \\ w^{(1)}_{\mathrm{eq},L-1}(p) & \cdots & \cdots & w^{(1)}_{\mathrm{eq},2L-2}(p) \end{pmatrix}, \tag{D.20}$$

$$\mathbf{A}_{21}(p) = \begin{pmatrix} w^{(2)}_{\mathrm{eq},0}(p) & w^{(2)}_{\mathrm{eq},1}(p) & \cdots & w^{(2)}_{\mathrm{eq},L-1}(p) \\ w^{(2)}_{\mathrm{eq},1}(p) & w^{(2)}_{\mathrm{eq},2}(p) & \cdots & w^{(2)}_{\mathrm{eq},L}(p) \\ \vdots & & & \vdots \\ w^{(2)}_{\mathrm{eq},L-1}(p) & \cdots & \cdots & w^{(2)}_{\mathrm{eq},2L-2}(p) \end{pmatrix}, \tag{D.21}$$

$$\mathbf{A}_{22}(p) = \begin{pmatrix} \tilde{w}^{(1)}_{\mathrm{eq},0}(p) & \tilde{w}^{(1)}_{\mathrm{eq},1}(p) & \cdots & \tilde{w}^{(1)}_{\mathrm{eq},L-1}(p) \\ \tilde{w}^{(1)}_{\mathrm{eq},-1}(p) & \tilde{w}^{(1)}_{\mathrm{eq},0}(p) & \cdots & \tilde{w}^{(1)}_{\mathrm{eq},L-2}(p) \\ \vdots & \ddots & \ddots & \vdots \\ \tilde{w}^{(1)}_{\mathrm{eq},-L+1}(p) & \cdots & \cdots & \tilde{w}^{(1)}_{\mathrm{eq},0}(p) \end{pmatrix}. \tag{D.22}$$

The coefficients $w^{(n)}_{\mathrm{eq},k}(p)$ come from the convolution of $\mathbf{r}_n(p)$ and $\mathbf{w}_{\mathrm{eq}}$ and are given by

$$w^{(n)}_{\mathrm{eq},k}(p) = \sum_{u=0}^{2L-2} w_{\mathrm{eq},u} r_{n,k-u}(p). \tag{D.23}$$

The coefficients $\tilde{w}^{(n)}_{\mathrm{eq},k}(p)$ are similarly defined, as follows: Let us introduce the response $\tilde{w}_{\mathrm{eq}}$, which is a shifted version of w_{eq},

$$\tilde{w}_{\mathrm{eq},k} = w_{\mathrm{eq},k+2d}. \tag{D.24}$$

Then, according to (D.4), we have

$$\tilde{w}_{\mathrm{eq},k} = \delta_k - \sum_{u=0}^{L-1} h_{12,u+d} h_{21,k+d-u}, \tag{D.25}$$

and we define $\tilde{w}^{(n)}_{\mathrm{eq},k}(p)$ as the convolution of $\tilde{w}_{\mathrm{eq},k}$ and $r_{n,k}(p)$:

$$\tilde{w}^{(n)}_{\mathrm{eq},k}(p) = \sum_{u=0}^{2L-2} \tilde{w}_{\mathrm{eq},u} r_{n,k-u}(p). \tag{D.26}$$

Now, we consider the situation where each source is silent over a certain time block T, say, $T_k = [kL - 3L + 3, kL]$, while the other sources are not silent over this interval:

$$\forall n = 1,\ldots,N \quad \exists k \text{ such that } \mathbf{s}_n(kL) = \mathbf{0} \text{ and } \forall m \neq n, \ \mathbf{s}_m(kL) \neq \mathbf{0}. \tag{D.27}$$

Without loss of generality, suppose s_1 is silent on the first time block $\mathcal{T}_1$. Let $\sigma_1^2(L)$ tend to zero in (D.17). One finds that the second term on the right-hand side of (D.17), namely

$$\frac{1}{\sigma_1^2(kL)}\mathbf{A}_{11}(kL)\varepsilon_{12}(n), \tag{D.28}$$

tends to $+\infty$. By contrast, the last term of (D.17), namely

$$\frac{1}{\sigma_1^2(kL)}\mathbf{A}_{12}(kL)\varepsilon_{21}(n), \tag{D.29}$$

remains bounded. Since we are interested in the *direction* of the update and not in its norm, we can consider a small step-size μ proportional to $\sigma_1^2(L)$ so that the limit of the second term on the right-hand side (D.17) exists. Then (D.29) vanishes at this limit, as well as the contributions of the arguments of the sum in (D.17) for $k = 2, \ldots, K$. One obtains the following relation on $\varepsilon_{12}(n)$:

$$\varepsilon_{12}(n + 1) = (\mathbf{I} - \tilde{\mu}\mathbf{A}_{11}(L))\,\varepsilon_{12}(n), \tag{D.30}$$

where $\tilde{\mu}\sigma_1^2(L) = \mu$. The same reasoning for source s_2 silent at the second time block $\mathcal{T}_2$ leads to a similar equation for $\varepsilon_{21}(n)$:

$$\varepsilon_{21}(n + 1) = (\mathbf{I} - \tilde{\mu}\mathbf{A}_{22}(2L))\,\varepsilon_{12}(n). \tag{D.31}$$

Therefore, the local stability depends on the positiveness of the eigenvalues of $\mathbf{A}_{11}(L)$ and $\mathbf{A}_{22}(2L)$.

D.3 Local Stability Conditions

Causal mixing, white sources

Suppose that the source s_2 is white in $\mathcal{T}_1$, then its correlation function $r_{2,\tau}$ is a Dirac impulse, $r_{2,\tau} = \delta(\tau)$. If additionally $d = 0$, then the definitions (D.23) and (D.24) yield

$$\tilde{w}_{\text{eq}}^{(n)}(p) = w_{\text{eq}}^{(n)}(p) = w_{\text{eq}}. \tag{D.32}$$

If $d = 0$, i.e., if the mixing is causal, then $w_{\text{eq},k} = 0$ for $k < 0$. Consequently, according to (D.19), the matrices $\mathbf{A}_{11}(L)$ and $\mathbf{A}_{22}(2L)$ in (D.30) and (D.31) become upper triangular. The eigenvalues are the diagonal elements, $w_{\text{eq},0} = 1 - h_{12,0}h_{21,0}$. Therefore, a *necessary and sufficient* condition for the local stability is

$$\boxed{1 - h_{12,0}h_{21,0} > 0.} \tag{D.33}$$

General case

The matrix $\mathbf{A}_{11}(L)$ in (D.30) is a Toeplitz matrix and can be made circulant using the elements of its first row and of its first column. Let us define

$$\mathbf{c}_1 = \left(\tilde{w}^{(2)}_{\text{eq},0}(L), \ldots, \tilde{w}^{(2)}_{\text{eq},-L+1}(L), \tilde{w}^{(2)}_{\text{eq},L-1}(L), \ldots, \tilde{w}^{(2)}_{\text{eq},1}(L) \right)^{\mathrm{T}}. \quad (D.34)$$

The time argument L appears because the local behavior is dominated by the source statistics at time block $\mathcal{T}_1$, as in (D.30). We denote the $2L - 1 \times 2L - 1$ circulant matrix with first column $\mathbf{c}_1$ by $\mathbf{C}$. Similarly, we double the size of ε_{12}:

$$\tilde{\varepsilon}_{12} = \begin{bmatrix} \varepsilon_{12} \\ \mathbf{0}_{L-1 \times 1} \end{bmatrix} \quad (D.35)$$

and we define the projection

$$\mathbf{\Omega} = \begin{bmatrix} \mathbf{I}_{L \times L} & \mathbf{0}_{L \times L-1} \\ \mathbf{0}_{L-1 \times L} & \mathbf{0}_{L-1 \times L-1} \end{bmatrix}. \quad (D.36)$$

Then (D.30) becomes

$$\tilde{\varepsilon}_{12}(n+1) = \mathbf{\Omega} \left(\mathbf{I} - \tilde{\mu}\mathbf{C} \right) \tilde{\varepsilon}_{12}(n). \quad (D.37)$$

Since $\mathbf{C}$ is circulant, it is diagonalized by the discrete Fourier transform (DFT) matrix $\mathbf{F}$, that is, $\mathbf{C} = \mathbf{F}\mathbf{\Lambda}\mathbf{F}^H$. Let us define $\mathbf{F}_0 = \mathbf{\Omega}\mathbf{F}$. We have $\mathbf{F}_0\mathbf{F}_0^H = \mathbf{\Omega}$ and we can transform (D.37) into

$$\tilde{\varepsilon}_{12}(n+1) = \mathbf{\Omega} \left(\mathbf{\Omega} - \tilde{\mu}\mathbf{F}_0\mathbf{\Lambda}\mathbf{F}_0^H \right) \tilde{\varepsilon}_{12}(n), \quad (D.38)$$

$$= \mathbf{\Omega}\mathbf{F}_0 \left(\mathbf{I} - \tilde{\mu}\mathbf{\Lambda} \right) \mathbf{F}_0^H \tilde{\varepsilon}_{12}(n). \quad (D.39)$$

This yields

$$\mathbf{F}_0^H \tilde{\varepsilon}_{12}(n+1) = \mathbf{\Omega} \left(\mathbf{I} - \tilde{\mu}\mathbf{\Lambda} \right) \mathbf{F}_0^H \tilde{\varepsilon}_{12}(n). \quad (D.40)$$

Equation (D.40) shows that a *sufficient* condition for the local stability is that the diagonal elements of $\mathbf{\Lambda}$, i.e., the DFT values of $\tilde{w}^{(2)}_{\text{eq}}$, have positive real parts. Let us denote by $\widetilde{W}^{(2)}_{\text{eq}}(k)$ the kth frequency bin of the DFT of $\tilde{w}^{(2)}_{\text{eq}}$, for $k = 0, \ldots, 2L - 2$:

$$\widetilde{W}^{(2)}_{\text{eq}}(k) = \sum_{\tau=-L+1}^{L-1} \tilde{w}^{(2)}_{\text{eq},\tau}(L) \, e^{2i\pi\tau k/(2L-1)}. \quad (D.41)$$

We similarly define $\widetilde{W}_{\text{eq}}(k)$ and $R_2(k)$, the kth frequency bin of $\tilde{w}_{\text{eq}}$ and $r_2(L)$ (the correlation function of the source s_2 at time block $\mathcal{T}_1$), respectively. According to (D.26), $\widetilde{W}^{(2)}_{\text{eq}}(k)$ can be factorized as

$$\widetilde{W}_{\text{eq}}^{(2)}(k) = \widetilde{W}_{\text{eq}}(k)R_2(k), \tag{D.42}$$

provided the response $\tilde{w}_{\text{eq}}$ of length L_w and the correlation function r_2 of length L_r fulfill $L_w + L_r - 1 \leq 2L - 1$ (which is true if the filters h_{12} and h_{21} are short enough). Since r_2 is symmetric, its DFT $R_2(k)$ is real-valued. It is further assumed that $R_2(k)$ is positive. According to (D.25), we have

$$\widetilde{W}_{\text{eq}}(k) = 1 - e^{-4i\pi dk/(2L-1)}H_{12}(k)H_{21}(k). \tag{D.43}$$

The positiveness of $\tilde{W}_{\text{eq}}^{(2)}(k)$ is obtained if $\left|e^{-4i\pi dk/(2L-1)}H_{12}(k)H_{21}(k)\right| < 1$, that is, if

$$\boxed{|H_{12}(k)H_{21}(k)| < 1} \tag{D.44}$$

which is the result used in Chap. 7. Note that the same reasoning applied to $\varepsilon_{21}(n)$ leads to the same sufficient local stability conditions.

E

Notations

Conventions

x	real scalar
$\mathbf{x}$	vector
$\mathbf{X}$	matrix or vector for a frequency-domain variable
$\mathbf{X}^H$	matrix Hermitian (complex conjugate) transpose
$\mathbf{X}^T$	matrix transpose
$\mathbf{X}^{-T}$	inverse of $\mathbf{X}^T$
$\triangleq$	definition (as opposed to assignment or assertion)

Abbreviations and acronyms

s.t.	subject to
AIC	Adaptive Interference Canceler
BGSC	Blind Generalized Sidelobe Canceler
BM	Blocking Matrix
BSS	Blind Source Separation
DOA	Direction of Arrival
DTD	Double-Talk Detector
DFT	Discrete Fourier Transform
DTFT	Discrete-Time Fourier Transform
FD-SOS	Frequency-Domain Second-Order Statistics BSS algorithm
FD-HOS	Frequency-Domain Higher-Order Statistics BSS algorithm
FFT	Fast Fourier Transform
FIR	Finite Impulse Response
GSC	Generalized Sidelobe Canceler
GSD	Generalized Sidelobe Decorrelator
IFFT	Inverse Fast Fourier Transform
ILMS	Implicit LMS
LCMV	Linearly Constrained Minimum Variance
LMS	Least-Mean Square
LS	Least Square
LTI	Linear Time Invariant

MIMO	Multiple-Input Multiple-Output
MISO	Multiple-Input Single-Output
NG-SOS-BSS	Natural Gradient Second-Order Statistics Blind Source Separation algorithm
PSD	Power Spectral Density
QIC	Quadratic Inequality Constraint
RGSC	Robust Generalized Sidelobe Canceler
STFT	Short-Time Discrete Fourier Transform
SOS-BSS	Second-Order Statistics Blind Source Separation
ULA	Uniform Linear Array
WER	Word Error Rate

Scalar variables

a_{ILMS}	ILMS convergence contraction factor
a_{NLMS}	NLMS convergence contraction factor
a_{QIC}	constant upper limit in the quadratic inequality constraint
$b(p)$	target signal at the output when $\mathbf{a}(p) = \mathbf{a}_{\mathrm{opt}}(p)$
b_{ILMS}	ILMS divergence contraction factor
b_{NLMS}	NLMS divergence contraction factor
d	number of acausal coefficients which are treated as acausal in the convolution operator $\star_d$
f_s	sampling frequency
$h_{mn,k}$	kth tap of the discrete-time acoustic channels from the nth source to the mth receiver
$h_{\mathbf{r},\boldsymbol{\theta}}$	continuous time acoustic channel from the source position $\boldsymbol{\theta}$ to the receiver position $\mathbf{r}$
p	discrete time
$s_n(p)$	nth source signal (speech source)
$s_{\boldsymbol{\theta}}(t)$	continuous time signal emitted at position $\boldsymbol{\theta}$
t	continuous time
$x(\mathbf{r}, t)$	continuous time signal received at position $\mathbf{r}$
$x_n(p)$	nth observed signal (microphone signal)
$w_{m,k}$	kth tap of the separation filter for the mth input of a MISO separation system
$w_{nm,k}$	kth tap of the separation filter for the mth input and nth output of a MIMO separation system
$x_0(p)$	target reference signal (fixed beamformer output signal)
$x_{B,m'}(p)$	blocking matrix m'th output signal
$x_m(p)$	mth received signal (microphone signal)
$x_{\mathrm{int},m}(p)$	contribution of the interference signal in the input signal $x_m(p)$
$x_{\mathrm{sig},m}(p)$	contribution of the desired signal in the input signal $x_m(p)$
$y(p)$	output signal for MISO separation system
$y_n(p)$	nth output signal for MIMO separation system
$y_{\mathrm{int}}(p)$	contribution of the interferences at the output

C	dimension of the constraint for LCMV beamformer
C_A	complexity of an addition or subtraction
C^{batch}	complexity of a batch algorithm
$C^{\text{block-wise}}$	complexity of a block-wise batch algorithm
$C_{\text{conv}}^{L_{\text{FFT}}}$	complexity for convolution in the DFT domain with FFT of length L_{FFT}
C_M	complexity of a multiplication or division
D	delay in the fixed beamformer path
$G_{\mathbf{w}}(\omega, \boldsymbol{\theta})$	space–frequency response
$H(\mathbf{y})$	entropy of $\mathbf{y}$
$H_{mn}(k)$	in the local stability analysis $H_{mn}(k)$ is the DFT of h_{mn} padded with $L-1$ zeros at frequency bin $(-k)$
$I(\mathbf{y})$	mutual information of $\mathbf{y}$
$\text{IR}(p)$	reduction of the interference signal level for compact arrays
$\text{IR}_{[t_0, t_1]}$	reduction of the interference signal level, averaged between times t_0 and t_1, for compact arrays
$\text{IR}^{\text{d}}(p)$	reduction of the interference signal level for distributed arrays
$\text{IR}^{\text{d}}_{[t_0, t_1]}$	reduction of the interference signal level, averaged between times t_0 and t_1, for distributed arrays
J	cost function (or "criterion")
$J\|_S$	restriction of J to the Sylvester subspace S
$J_{\text{LS}}(\mathbf{w})$	least-square cost function
J_{LMS}	least-mean-square criterion
$J_{\text{BSS}}(\mathbf{W})$	BSS cost function
$J_{\text{BSS,LS}}(\mathbf{W})$	least-square BSS cost function
$J_{\text{geo}}(\mathbf{W})$	geometric cost function
K	number of jointly diagonalized output correlation matrices in SOS-BSS
L	length of the separation filters
L_{m}	length of the mixing channels
M	number of microphones
M'	number of interference reference signals
N	number of sources
N_{iter}	number of iterations in the gradient descent
Q	quality measure to determine the constants μ_{NLMS}, μ_0, and a_{QIC}
$Q_{[t_0, t_1]}$	signal-to-interference ratio improvement, averaged between times t_0 and t_1, for compact arrays
$Q^{\text{d}}_{[t_0, t_1]}$	signal-to-interference ratio improvement, averaged between times t_0 and t_1, for distributed arrays
$\text{SIR}^{\text{d}}_{\text{imp}}(p)$	signal-to-interference ratio improvement for distributed arrays

$\mathrm{SIR_{imp}}(p)$ signal-to-interference ratio improvement for compact arrays
$\mathrm{SR}(p)$ reduction of the target signal level for compact arrays
$\mathrm{SR}_{[t_0,t_1]}$ reduction of the target signal level, averaged between times t_0 and t_1, for compact arrays
$\mathrm{SR^d}(p)$ reduction of the target signal level for distributed arrays
$\mathrm{SR}^{\mathrm{d}}_{[t_0,t_1]}$ reduction of the target signal level, averaged between times t_0 and t_1, for distributed arrays
T number of samples

Vectors and matrices

$\mathbf{a}(p)$ adaptive interference canceler
$\mathbf{a}_{\mathrm{opt}}(p)$ optimal adaptive interference canceler (Wiener solution)
$\mathbf{c}$ $C \times 1$ response vector
$\mathbf{c}_{\mathrm{int}}$ global interference–output response for $MISO$ separation systems
$\mathbf{c}_n$ global nth source–output response for $MISO$ separation systems
$\mathbf{d}(p)$ contribution of the desired signal in $\mathbf{x}(p)$
$\mathbf{d}(\theta)$ $M(2L - 1) \times 1$ vector used to define the spatial response $\mathbf{g}(\mathbf{W}_n, \theta)$
$\mathbf{g}(\mathbf{W}_n, \theta)$ spatial response defined with the Sylvester matrix $\mathbf{W}_n$
$\mathbf{g}(\mathbf{w}, \theta)$ spatial response of the MISO separation system $\mathbf{w}$ for the DOA θ
$\mathbf{g}(\mathbf{w}, \boldsymbol{\theta})$ spatial response of the MISO separation system $\mathbf{w}$ for the 3D position $\boldsymbol{\theta}$
$\mathbf{h}_{mn}$ $L_{\mathrm{m}} \times 1$ mixing channel vector for the nth source and mth receiver
$\mathbf{m}(p)$ mismatch between the actual adaptive interference canceler and the Wiener solution
$\mathbf{n}(p)$ contribution of the local interference signal in $\mathbf{x}(p)$
$\mathbf{n}^{(\mathrm{road})}(p)$ contribution of the road noise signal $\mathbf{x}(p)$
$\mathbf{s}_{\mathrm{int}}(p)$ $(N - 1)(L + L_{\mathrm{m}} - 1) \times 1$ interference source signal vector
$\mathbf{s}_n(p)$ $L + L_{\mathrm{m}} - 1 \times 1$ nth source vector
$\mathbf{w}$ $ML \times 1$ multichannel filter vector for MISO separation system
$\mathbf{w}_0$ fixed beamformer in a GSC beamformer
$\mathbf{w}_m$ $L \times 1$ separation filter for the mth input of a MISO separation system
$\mathbf{w}_{nm}$ $L \times 1$ separation filter vector for MIMO separation system
$\mathbf{x}(p)$ $ML \times 1$ multichannel input signal vector for MISO separation system
$\mathbf{x}(p)$ $M(2L - 1) \times 1$ input vector for a MIMO separation system
$\mathbf{x}_B(p)$ interferer reference signal (blocking matrix output signal)
$\mathbf{x}_m(p)$ $L \times 1$ time-reversed mth input signal vector

$\mathbf{y}(p)$ $NL \times 1$ vector representing the outputs of a MIMO separation system

$\mathbf{y}_{[i,\ldots,j]}(p)$ $(j - i + 1)L \times 1$ vector representing the output signals $\mathbf{y}_i(p), \ldots, \mathbf{y}_j(p)$

$\mathbf{y}_n(p)$ $L \times 1$ vector representing the nth output signal of a MIMO separation system

$\mathbf{B}$ blocking matrix in a GSC beamformer

$\mathbf{C}$ $ML \times C$ constraint matrix, or block Sylvester mixing matrix of size $NL \times N(L_m + 2L - 2)$ representing the global source–output MIMO system

$\mathbf{C}_{nn'}$ Sylvester mixing matrix of size $L \times (L_m + 2L - 2)$ representing a source–output channel

$\mathbf{D}$ matrix used to define a geometric constraint

$\mathbf{D}(\omega, \boldsymbol{\theta})$ steering vector for the source position $\boldsymbol{\theta}$

$\mathbf{G}$ matrix defining a general geometric constraint

$\mathbf{G}_0$ constraint of zero spatial response at the interference DOAs

$\mathbf{G}_1$ constraint of unit spatial response at the target position

$\mathbf{G}_2$ constraint of unit spatial response at the target position and of zero spatial response at the interference DOA

$\mathbf{H}$ block Sylvester mixing matrix of size $M(2L-1) \times N(L_m + 2L - 2)$

$\mathbf{H}_{int}$ $(N-1)(L + L_m - 1) \times ML$ interference mixing matrix

$\mathbf{H}_{mn}$ $L + L_m - 1 \times L$ matrix representing the mixing channel h_{mn} used in $\mathbf{H}_{int}$

$\mathbf{H}_{mn}$ $(2L-1) \times (L_m + 2L - 2)$ Sylvester matrix representing the mixing channel $\mathbf{h}_{nm}$

$\mathbf{I}$ identity matrix

$\mathbf{R}_{int}(p)$ $\mathbf{s}_{int}(p)$ correlation matrix

$\mathbf{R}_{\mathbf{xx}}(p)$ $ML \times ML$ correlation matrix for the input signal vector $\mathbf{x}(p)$

$\mathbf{R}_{\mathbf{x}_B\mathbf{x}_B}(p)$ correlation matrix for $\mathbf{x}_B(p)$

$\widehat{\mathbf{R}}_{\mathbf{x}_B\mathbf{x}_B}(p)$ estimation of $\mathbf{R}_{\mathbf{x}_B\mathbf{x}_B}(p)$

$\mathbf{R}_{\mathbf{yy}}(p)$ correlation matrix for the output vector $\mathbf{y}(p)$

$\widehat{\mathbf{R}}_{\mathbf{yy}}(p)$ estimated correlation matrix for the output vector $\mathbf{y}(p)$

$\mathbf{R}_{\mathbf{y}_n\mathbf{y}_n}(p)$ correlation matrix for the output vector $\mathbf{y}_n(p)$ in the nth channel

$\widehat{\mathbf{S}}_{\mathbf{y}_i\mathbf{y}_j}^{(n)}$ regularized output correlation matrix

$\mathbf{W}$ block Sylvester matrix representing the entire MIMO separation system

$\mathbf{W}(\omega)$ DTFT of the separation system $\mathbf{w}$

$\mathbf{W}(p)$ in sample-wise adaptive mode, the separation matrix $\mathbf{W}$ at time p

$\mathbf{W}(n)$ in batch mode, the separation matrix $\mathbf{W}$ at the nth iteration

$\mathbf{W}_n$ nth row of $\mathbf{W}$

$\mathbf{W}(n, p)$ in block-wise batch adaptive mode, the separation matrix $\mathbf{W}$ at the nth iteration in the block at time p

$\Delta \mathbf{W}(n)$	update for the matrix $\mathbf{W}(n)$, that is, $\mathbf{W}(n+1) = \mathbf{W}(n) - \mu(n)\Delta\mathbf{W}(n)$
$\Delta\mathbf{W}^{(\mathrm{BSS})}$	BSS update
$\Delta\mathbf{W}^{(\mathrm{geo})}$	geometric update
$\mathbf{W}_{nm}$	$L \times (2L-1)$ Sylvester matrix representing the separation filter $\mathbf{w}_{nm}$
$\mathbf{W}_{\mathrm{opt}}$	optimum separation matrix (equilibrium point)
$\mathbf{W}_{\mathrm{opt}}$	optimum separation matrix (defined as the minimum of $\xi_{BSS}(\mathbf{W})$
$\widehat{\mathbf{W}}_{\mathrm{opt}}$	estimate of $\mathbf{W}_{\mathrm{opt}}$
$\overline{\mathbf{W}}$	multichannel z-transform of the Sylvester matrix $\mathbf{W}$

Functions and operators

$\mathrm{bdiag}\mathbf{A}$	for a block matrix $\mathbf{A}$, operator that sets the off-diagonal submatrices to $\mathbf{0}$
$\mathrm{bdiag}_{1,[2,\ldots,M]}\mathbf{A}$	operator setting the off-diagonal submatrices $\mathbf{A}_{1m}$ and $\mathbf{A}_{m1}$ for $m = 2,\ldots,M$ to $\mathbf{0}$
$\mathrm{boff}\mathbf{A}$	for a block matrix $\mathbf{A}$, operator that sets the diagonal submatrices to $\mathbf{0}$
$\det(\mathbf{A})$	determinant of the matrix $\mathbf{A}$
$D_g^{(\mathbf{W}_{\mathrm{opt}})}(\varepsilon(n))$	derivative of g at point $\mathbf{W}_{\mathrm{opt}}$, as a linear function of a small deviation
$\mathbf{E}\{\}$	expectation operator (ensemble average)
$\widehat{\mathbf{E}}\{\}$	estimation of $\mathbf{E}\{\}$
$f^{(C)}(\cdot)$	Cth composition of a function f
$S(\mathbf{A})$	operator which transforms a general matrix $\mathbf{A}$ into a Sylvester matrix by summing the redundant terms
$S_d(\mathbf{A})$	operator which transforms a general matrix $\mathbf{A}$ into a Sylvester matrix using the dth row as reference
$S^L(\mathbf{A})$	operator which transforms a general matrix $\mathbf{A}$ into a Sylvester matrix using the Lth column as reference
$S_{\mathrm{approx}}(\mathbf{A})$	generic approximation of $S(\mathbf{A})$ representing either $S_d(\mathbf{A})$ or $S^L(\mathbf{A})$
$\mathrm{tr}(\mathbf{A})$	trace of the matrix $\mathbf{A}$
$\|\mathbf{x}\|$	$\sqrt{\mathbf{x}^H\mathbf{x}}$ (vector norm)
$\|\mathbf{X}\|$	$\sqrt{\mathrm{tr}(\mathbf{X}^H\mathbf{X})}$ (matrix norm)
$*$	convolution operator
$\star_d$	convolution where d coefficients are treated as acausal and where the result is truncated on L coefficients
$[\overline{\mathbf{B}}]_{\mathcal{S}}(z)$	operator that truncates a z-transform $\overline{\mathbf{B}}$ to a support $\mathcal{S} \subset [-L+1, L-1]$
$\langle \cdot, \cdot \rangle$	scalar product associated to the Euclidean metric
$\lceil x \rceil$	smallest integer larger than x
$\lfloor x \rfloor$	largest integer smaller than x

Sets

$\mathbb{N}$	natural numbers
$\mathbb{Z}$	integers
$\mathbb{R}$	real numbers
$\mathbb{C}$	complex numbers
$\boldsymbol{S}$	space of the block Sylvester matrices of size $NL \times M(2L-1)$
$\overline{\boldsymbol{S}}$	set of the single-sided z-transforms of length L
$\boldsymbol{T}$	space of the block Toeplitz matrices of size $NL \times NL$
$\overline{\boldsymbol{T}}$	set of the multichannel two-sided z-transforms

Greek letters

α	regularization parameter
$\alpha(p)$	general contraction factor
β	βL is the number of samples in the most recent input block for block-wise batch algorithms
δ	fixed regularization term
$\boldsymbol{\delta}_d$	vector representing a delay of d taps
Δ	interelement spacing for a uniform linear array
$\varepsilon_{\text{mismatch}}$	factor representing the interferer signal power at the beamformer output
$\varepsilon_{\text{leakage}}$	factor representing the amount of target leakage into the interference reference
$\boldsymbol{\varepsilon}(n)$	small deviation around the equilibrium point $\mathbf{W}_{\text{opt}}$
λ	geometric weight
λ_i	eigenvalue of $\mathbf{R}_{\mathbf{yy}}$
$\tilde{\lambda}_i$	error on the eigenvalue λ_i
$\lambda_i^{(n)}$	eigenvalue of $\mathbf{R}_{\mathbf{yy}}^{(n)}$, the output correlation matrix at iteration n
$\lambda_{\max}$	largest eigenvalue of $\widehat{\mathbf{R}}_{\mathbf{x}_B\mathbf{x}_B}$ or $\mathbf{R}_{\mathbf{x}_B\mathbf{x}_B}$
μ	step-size
$\tilde{\mu}$	normalized step-size
μ_0	step-size for the ILMS algorithm
μ_{LMS}	step-size for the LMS algorithm
$\mu_{\max}$	maximal step-size for the stability of ILMS
μ_{NLMS}	step-size for the NLMS algorithm
θ	direction of arrival
$\boldsymbol{\theta}$	3D source position
θ_1	direction of arrival for the source of interest
σ_1^2	variance of the target signal at the beamformer output
σ_2^2	variance of the interference signal at the interference reference
$\tau_{m,\boldsymbol{\theta}}$	delay needed for a sound wave emitted at $\boldsymbol{\theta}$ to travel from the mth sensor to the next
ω	angular frequency

$\xi_{BSS}(\mathbf{W})$ BSS cost function involving the expectation operator
$\xi_{LMS}(\mathbf{w})$ least-mean-square cost function involving the expectation operator

Special symbols

$\diamond$ symbol denoting an unconstrained spatial response

References

1. F. Abrard, Y. Deville and P. White. From blind source separation to blind source cancellation in the under-determined case: a new approach based on time-frequency analysis. *Proceedings of the International Symposium on Independent Component Analysis and Blind Signal Separation (ICA2001)*, San Diego, USA, December 2001.
2. S. Affes and Y. Grenier. A signal subspace tracking algorithm for microphone array processing of speech. *IEEE Transactions on Speech and Audio Processing*, Vol. 5, No. 5, pp. 425–437, September 1997.
3. R. Aichner, S. Araki, S. Makino, T. Nishikawa and H. Saruwatari. Time-domain blind source separation of non-stationary convolved signals with utilization of geometric beamforming. *Proceedings of the International workshop on Neural Networks for Signal Processing*, Martigny, Switzerland, 2002.
4. R. Aichner, H. Buchner, F. Yan and W. Kellermann. Real-time convolutive blind source separation based on a broadband approach. *Proceedings of the 5th International Symposium on Independent Component Analysis and Blind Signal Separation (ICA2004)*, pp. 840–848, Granada, Spain, September 2004.
5. R. Aichner, H. Buchner and W. Kellermann. On the causality problem in time-domain blind source separation and deconvolution algorithms. *Proceedings of the IEEE Internatinal Conference on Acoustics, Speech, and Signal Processing (ICASSP2005)*, Philadelphia, USA, March 2005.
6. S. Amari, S.C. Douglas, A. Cichocki and H.H. Yang. Multichannel blind deconvolution and equalization using the natural gradient. *Proceedings of the IEEE Workshop on Signal Processing Advances in Wireless Communications*, pp. 101–104, Paris, France, April 1997.
7. S. Amari. Natural gradient works efficiently in learning. *Neural Computation*, Vol. 10, pp. 251–276, 1998.
8. S. Amari, T. Chen and A. Cichocki. Nonholonomic orthogonal learning algorithm for blind source separation. *Neural Computation*, Vol. 12, pp. 1463–1483, 2000.
9. S. Araki, S. Makino, R. Mukai, T. Nishikawa and H. Saruwatari. The fundamental limitation of frequency-domain blind source separation for convolutive mixtures of speech. *Proceedings of the IEEE International Conference on Acoustics, Speech, and Signal Processing (ICASSP2001)*, pp. 2737–2740, Salt Lake City, USA, May 2001.

10. S. Araki, S. Makino, R. Mukai, Y. Hinamoto, T. Nishikawa and H. Saruwatari. Equivalence between frequency domain blind source separation and frequency domain adaptive beamforming. *Proceedings of the IEEE International Conference on Acoustics, Speech, and Signal Processing (ICASSP2002)*, Student Forum, Orlando, USA, May 2002.

11. S. Araki, S. Makino, Y. Hinamoto, R. Mukai, T. Nishikawa and H. Saruwatari. Equivalence between frequency-domain blind source separation and frequency-domain adaptive beamforming for convolutive mixtures. *EURASIP Journal on Applied Signal Processing*, Vol. 2003, No. 11, pp. 1157–1166, October 2003.

12. F. Asano, Y. Motomura, H. Asoh and T. Matsui. Effect of PCA filter in blind source separation. *Proceedings of the 2nd International Symposium on Independent Component Analysis and Blind Signal Separation (ICA2000)*, pp. 57–62, Helsinki, Finland, June 2000.

13. A.J. Bell and T.J. Sejnowski, An information-maximization approach to blind separation and blind deconvolution. *Neural Computation*, Vol. 7, pp. 1129–1159, 1995.

14. J. Bitzer. Mehrkanalige Geräuschunterdrückungssysteme – eine vergleichende Analyse. Dissertation, Arbeitsbereich Nachrichtentechnik, Universität Bremen, Mai 2002 (in German).

15. J. Bourgeois. A clustering approach to on-line audio source separation. *Proceedings of the European Conference on Speech and Language Processing (EUROSPEECH2003)*, pp. 1745–1748, Geneva, Switzerland, September 2003.

16. J. Bourgeois. An LMS viewpoint on the local stability of second order blind source separation. *Proceedings of the IEEE Workshop on Statistical Signal Processing*, Bordeaux, France, July 2005.

17. M.S. Brandstein and D.B. Ward, eds. *Microphone Arrays: Signal Processing Techniques and Applications*. Springer-Verlag, Berlin, 2001.

18. H. Buchner, R. Aichner and W. Kellermann. Blind source separation for convolutive mixtures: a unified treatment. In *Audio signal processing for next-generation multimedia communication systems*, Y. Huang and J. Benesty, eds., pp. 255–293. Kluwer Academic Publishers, Boston, February 2004.

19. H. Buchner, R. Aichner and W. Kellermann. Relation between blind system identification and convolutive blind source separation. *Conference Record of the Joint Workshop for Hands-Free Speech Communication and Microphone Arrays*, Piscataway, NJ, USA, March 2005

20. H. Buchner, R. Aichner and W. Kellermann. A generalization of blind source separation algorithms for convolutive mixtures based on second-order statistics. *IEEE Transactions on Acoustics, Speech, and Signal Processing*, Vol. 13, No. 1, January 2005.

21. J.F. Cardoso and A. Souloumiac. Blind beamforming for non-Gaussian signals. *Proceedings of the IEEE*, Vol. 140, No. 6, pp. 362–370, December 1993.

22. J.F. Cardoso and B. Laheld. Equivariant adaptive source separation. *IEEE Transactions on Signal Processing*, Vol. 44, No. 12, pp. 3017–3029, 1996.

23. J.F. Cardoso. Blind signal separation: statistical principles. *Proceedings of the IEEE*, Vol. 9, No. 10, pp. 2009–2025, October 1998.

24. J.F. Cardoso and D. T. Pham. Separation of non-stationary sources. Algorithms and performance. In *Independent Components Analysis: Principles and Practice*, S. J. Roberts and R. M. Everson, eds., pp. 158–180. Cambridge University Press, Cambridge, MA, 2001.

25. A. Cichocki and S. Amari. *Adaptive Blind Signal and Image Processing.* Wiley, Chichester, UK, 2002.

26. N. Charkani and Y. Deville. Self-adaptive separation of convolutively mixed signals with a recursive structure. Part I : stability analysis and optimization of asymptotic behaviour. *Signal Processing*, Vol. 73, No. 3, pp. 225–254, 1999.

27. D. Van Compernolle. Switching adaptive filters for enhancing noisy and reverberant speech from microphone array recordings. *Proceedings of the IEEE International Conference on Acoustics, Speech, and Signal Processing*, Vol. 2, pp. 833–836, Albuquerque, NM, USA, April 1990.

28. R.T. Compton. The energy-inversion adaptive array: concept and performance. *IEEE Transactions on Aerospace and Electricity Systems*, Vol. AES-15, pp. 803–814, November 1979.

29. T. Cover and J. Thomas. *Elements of Information Theory.* John Wiley, New York, 1991.

30. H. Cox, R. M. Zeskind and M. M. Owen. Robust adaptive beamforming. *IEEE Transactions on Acoustics, Speech, and Signal Processing*, Vol. 35, No. 10, pp. 1365–1376, 1987.

31. S. Cruces and A. Cichocki. Globally convergent newton algorithms for blind decorrelation, *Proceedings of the 4th International Symposium on Independent Component Analysis and Blind Signal Separation* (ICA2003), pp. 421–426, Nara, Japan, April 2003.

32. S.C. Douglas and A. Cichocki. Neural networks for blind decorrelation of signals. *IEEE Transactions on Signal Processing*, Vol. 45, No. 11, pp. 2829–2842, November 1997.

33. S.C. Douglas and X. Sun. Convolutive blind separation of speech mixtures using the natural gradient. *Speech Communication*, Vol. 39 , pp. 65–78, January 2003.

34. S.C. Douglas, H. Sawada and S. Makino. On coefficient delay in natural gradient blind deconvolution and source separation algorithms. *Proceedings of the 5th International Symposium on Independent Component Analysis and Blind Signal Separation (ICA2004)*, pp. 634–642, Granada, Spain, September 2004.

35. C. Fancourt and L. Parra. The generalized sidelobe decorrelator. *Proceedings of the IEEE Workshop on the Applications of Signal Processing to Audio and Acoustics*, pp. 167–170, New Platz, USA, October 2001.

36. O.L. Frost III. An algorithm for linearly constrained adaptive array processing. *Proceedings of the IEEE*, Vol. 60, pp. 926–935, August 1972.

37. S. Gannot, D. Burshtein and E. Weinstein. Signal enhancement using beamforming and non-stationarity with application to speech. *IEEE Transactions on Signal Processing*, Vol. 49, No. 8, pp. 1614–1626, September 2001.

38. S. Van Gerven and D. Van Compernolle. Signal separation by symmetric adaptive decorrelation: stability, convergence, and uniqueness. *IEEE Transactions on Signal Processing*, Vol. 43, No. 7, July 1995.

39. S. Van Gerven. Adaptive noise cancellation and signal separation with applications to speech enhancement. PhD thesis, K.U.Leuven, ESAT, March 1996.

40. A. Gilloire and J.-P. Jullien. L'acoustique des salles dans les télécommunications. *L'écho des recherches*, No. 127, pp. 43–54, 1987 (in French).

41. J. Greenberg. Modified LMS Algorithm for speech processing with an adaptive noise canceller. *IEEE Transactions on Speech and Audio Processing*, Vol. 6, No. 4, pp. 338–351, July 1998.

42. L.J. Griffiths and C.W. Jim. An alternative approach to linearly constrained adaptive beamforming. *IEEE Transactions on Antennas and Propagation*, Vol. AP-30, pp. 27–34, January 1982.

43. A. Gruenstein, J. Niekrasz and M. Purver. Meeting structure annotation: data and tools. *Proceedings of the 6th SIGdial Workshop on Discourse and Dialogue*, Lisbon, Portugal, 2005.

44. E. Hänsler and G. U. Schmidt. *Acoustic Echo and Noise Control: A Practical Approach*. Wiley, 2004.

45. S. Haykin. *Adaptive Filter Theory*. Second Edition, Prentice Hall, NJ, 1996.

46. S. Haykin. Introduction. In *Unsupervised Adaptive Filtering. Vol. 1: Blind Source Separation*, S. Haykin, ed., pp. 1–12, John Wiley, New York, 2000.

47. W. Herbordt and W. Kellermann. Analysis of blocking matrices for generalized sidelobe cancellers for non-stationary broadband signals. *Proceedings of the IEEE International Conference on Acoustics, Speech, and Signal Processing (ICASSP2002)*, Student Forum, Orlando, USA, May 2002.

48. W. Herbordt and W. Kellermann. Adaptive beamforming for audio signal acquisition. In *Adaptive Signal Processing: Applications to Real-World Problems*, J. Benesty and Y. Huang, eds., pp. 155–194, Springer, Berlin, 2003.

49. W. Herbordt, T. Trini and W. Kellermann. Robust spatial estimation of the signal-to-interference ratio for non-stationary mixtures. *Proceedings of the International Conference on Acoustic Echo and Noise Control*, pp. 247–250, Kyoto, Japan, September 2003.

50. W. Herbordt. Combination of robust adaptive beamforming with acoustic echo cancellation for acoustic human/machine interfaces. PhD thesis, Universität Erlangen-Nürnberg, 2003.

51. Markus Hofbauer. On the FIR inversion of an acoustical convolutive mixing system: properties and limitations. *Proceedings of the 5th International Symposium on Independent Component Analysis and Blind Signal Separation (ICA2004)*, pp. 840–848, Granada, Spain, September 2004.

52. T.P. von Hoff. On the convergence of blind separation and deconvolution. PhD thesis, ETH Zurich, Hartung-Gorre Verlag, Konstanz, 2000.

53. O. Hoshuyama and A. Sugiyama. A robust adaptive beamformer for microphone arrays with a blocking matrix using constrained adaptive filters. *Proceedings of the IEEE International Conference on Acoustics, Speech and Signal Processing*, Atlanta, GA, pp. 925–928, May 1996.

54. O. Hoshuyama, B. Begasse, A. Sugiyama and A. Hirano. A real time robust adaptive microphone array controlled by an SNR estimate. *Proceedings of the IEEE International Conference on Acoustics, Speech and Signal Processing*, Vol. 6, pp. 3605–3608, 1998.

55. O. Hoshuyama and A. Sugiyama. Robust adaptive beamforming. In M. Brandstein and D. Wards, eds., *Microphone Arrays, Signal Processing Techniques and Applications.*, pp. 87–109. Springer, Berlin, 2001.

56. A. Hyvärinen, J. Karhunen and E. Oja. *Independent Component Analysis*. John Wiley, New York, 2001.

57. M. Z. Ikram and D. R. Morgan. Exploring permutation inconsistency in blind separation of speech signals in a reverberant environment. *Proceedings of the IEEE International Conference on Acoustics, Speech and Signal Processing*, Istanbul, Turkey, pp. 1041–1044, June 2000.

58. M. Z. Ikram and D. R. Morgan. A beamforming approach to permutation alignment for multichannel frequency-domain blind speech separation. *Proceedings of the IEEE International Conference on Acoustics, Speech, and Signal Processing (ICASSP2002)*, pp. 881–884, Orlando, USA, May 2002.

59. C. Jutten and H. Herault. A neuromimetic solution for the problem of source discrimination. *Traitement du Signal*, Vol. 5, No. 6, pp. 389–403, 1988.

60. K. D. Kammeyer and K. Kroschel. Digitale Signalverarbeitung. Teubner Studienbücher, Suttgart, 1998. (in German)

61. W. Kellermann and H. Buchner. Wideband algorithms versus narrowband algorithms for adaptive filtering in the DFT domain. *Proceedings of the Asilomar Conference on Signals, Systems, and Computers*, Pacific Grove, USA, November 2003.

62. Krause und Nesemann. Differenzengleichungen und diskrete dynamische Systeme. B.G.Teubner Stuttgart Leipzig, 1999 (in German).

63. S. Kurita, H. Saruwatari, S. Kajita, K. Takeda and F. Itakura. Evaluation of frequency-domain blind signal separation using directivity pattern under reverberant conditions. *Proceedings of the IEEE International Conference on Acoustics, Speech and Signal Processing*, Istanbul, Turkey, pp. 3140–3143, June 2000.

64. T.I. Laakso, V. Valimaki, M. Karhalainen and U.K. Laine. Splitting the unit delay. *IEEE Signal Processing Magazine*, Vol. 13, pp. 30–60, January 1996.

65. G. Lathoud, J. Bourgeois and J. Freudenberger. Sector-based detection for speech enhancement in cars. *EURASIP Journal on Applied Signal Processing, Special Issue on Microphone Arrays*, 2006.

66. J.S. Lim. *Speech Enhancement*. Prentice Hall, New Jersey, 1983.

67. S.Y. Low, S. Nordholm and R. Togneri. Convolutive blind signal separation with post-processing. *IEEE Transactions on Speech and Audio Processing*, Vol. 12, No. 5, pp. 320–327, September 2004.

68. A. Mader, H. Puder and G. U. Schmidt. Step-size control for acoustic echo cancellation. *Signal Processing*, Vol. 80, No. 9, pp. 1697–1719, September 2000.

69. V. Madisetti and D. B. Williams. *The Digital Signal Processing Handbook*. CRC Press, New York, 1998.

70. A. Mansour, C. Jutten and P. Loubaton. Adaptive subspace algorithm for blind separation of independent sources in convolutive mixtures. *IEEE Transactions on Signal Processing*, Vol. 48, No. 2, pp. 583–586, February 2000.

71. W. Minker, D. Bühler and L. Dybkjær. *Spoken Multimodal Human-Computer Dialogue in Mobile Environments*. Springer, Dordrecht (The Netherlands), 2005.

72. R. Mukai, S. Araki, H. Sawada and S. Makino. Removal of residual crosstalk components in blind source separation using LMS filters. *Proceedings of the International workshop on Neural Networks for Signal Processing*, pp. 435–444, Martigny, Switzerland, 2002.

73. R. Mukai, H. Sawada, S. Araki and S. Makino. Real-time blind source separation for moving speakers using blockwise ICA and residual crosstalk subtraction. *Proceedings of the 4th International Symposium on Independent Component Analysis and Blind Signal Separation* (ICA2003), pp. 421–426, Nara, Japan, April 2003.

74. T. Nishikawa, H. Saruwatari and K. Shikano. Blind source separation of acoustic signals based on multistage ICA combining frequency-domain ICA and time-domain ICA. *IEICE Transactions on Fundamentals*, Vol. E86-A, No. 4, pp. 846–858, April 2003.

75. L. Parra and C. Spence. Convolutive blind source separation of non-stationary sources. *IEEE Transactions on Speech and Audio Processing*, pp. 320–327, May 2000.

76. L. Parra and C. Spence. Online convolutive blind source separation of non-stationary sources. *Journal of VLSI Signal Processing*, Vol. 26, pp. 39–46, August 2000.

77. L. Parra and C. Alvino. Geometric source separation: merging convolutive source separation with geometric beamforming. *IEEE Transactions on Speech and Audio Processing*, Vol. 10, No. 6, pp. 352–362, September 2002.

78. S. Rickard, R. Balan and J. Rosca. Real-time time-frequency based blind source separation. *Proceedings of the International Symposium on Independent Component Analysis and Blind Signal Separation (ICA2001)*, pp. 421–426, San Diego, USA, December 2001.

79. E. Robledo-Arnuncio and B.H. Juang. Issues in frequency-domain blind source separation – a critical revisit. *Proceedings of the IEEE Internatinal Conference on Acoustics, Speech, and Signal Processing (ICASSP2005)*, pp. V281–284, Philadelphia, PA, USA, March 2005.

80. H. Saruwatari, K. Sawai, A. Lee, K. Shikano, A. Kaminuma and M. Sakata. Speech enhancement and recognition in car environment using blind source separation and subband elimination processing. *Proceedings of the 4th International Symposium on Independent Component Analysis and Blind Signal Separation (ICA2003)*, Nara, Japan, April 2003.

81. H. Sawada, R. Mukai, S. de la Kethulle de Ryhove, S. Araki and S. Makino. Spectral smoothing for frequency-domain blind source separation. *Proceedings of the International Workshop on Acoustic Echo and Noise Control (IWAENC2003)*, pp. 311–314, Kyoto, Japan, September 2003.

82. H. Sawada, R. Mukai, S. Araki and S. Makino. A robust and precise method for solving the permutation problem of frequency-domain blind source separation. *IEEE Transactions on Speech and Audio Processing*, Vol. 12, No. 5, pp. 530–538, September 2004.

83. R. Mukai, H. Sawada, S. Araki and S. Makino. Blind source separation and DOA estimation using small 3-D microphone array. *Proceedings of Joint Workshop on Hands-Free Speech Communication and Microphone Arrays (HSCMA 2005)*, pp. d9–10, Piscataway, March 2005.

84. I. Schwetz, G. Gruhler and K. Obermayer. Correlation and stationarity of speech radiation: consequences for linear multichannel filtering. *IEEE Transactions on Speech and Audio Processing*, Vol. 12, No. 5, pp. 460 – 467, September 2004.

85. C. E. Shannon. A mathematical theory of communication. *Bell System Technical Journal*, Vol. 27, pp. 379–423 and 623–656, July and October, 1948.

86. K. Torkolla. Blind separation for audio signals – are we there yet ? *Proceedings of the Workshop on Independent Component Analysis and Blind Signal Separation*, Aussois, France, January 1999.

87. S. Ukai, T. Takatani, T. Nishikawa and H. Saruwatari. Blind source separation combining SIMO-model-based ICA and adaptive beamforming. *Proceedings of the IEEE Internatinal Conference on Acoustics, Speech, and Signal Processing (ICASSP2005)*, pp. III85–88, Philadelphia, USA, March 2005.

88. B.D. Van Veen and K.M. Buckley. Beamforming: a versatile approach to spatial filtering. *IEEE Acoustics, Speech, Signal Processing Magazine*, Vol. 5, pp. 4–24, April 1988.

89. S.B. Weinstein. Echo cancellation in the telephone network. *IEEE Communications Society Magazine*, pp. 9–15, January 1977.

90. E. Weinstein, M. Feder and A.V. Oppenheim. Multi-channel signal separation by decorrelation. *IEEE Transactions on Speech and Audio Processing*, Vol. 1, No. 4, pp. 405–413, April 1993.

91. B. Widrow, K.M. Duvall, P.R. Gooch and W.C. Newmann. Signal cancellation phenomena in adaptive antennas: causes and cures. *IEEE Transactions on Acoustics, Speech and Signal Processing*, Vol. ASSP-30, pp. 469–478, May 1982.

92. B. Widrow and S. D. Stearn. *Adaptive Signal Processing*. Signal Processing Series. Prentice-Hall, Englewood Cliffs, 1985.

93. S. Winter, H. Sawada and S. Makino. Geometrical interpretation of the PCA subspace method for overdetermined blind source separation. *Proceedings of the 4th International Symposium on Independent Component Analysis and Blind Signal Separation* (ICA2003), pp. 775–780, Nara, Japan, April 2003.

94. R. Zelinski. A microphone array with adaptive post-filtering for noise reduction in reverberant rooms. *Proceedings of IEEE International Conference on Acoustics, Speech, and Signal Processing (ICASSP1988)*, pp. 2578–2581, New York, USA, March 2005.

95. L.Q. Zhang, A. Cichocki and S. Amari. Geometrical structures of FIR manifold and their application to multichannel blind deconvolution. *Proceedings of the IEEE Workshop on Neural Networks for Signal Processing*, pp. 303–312, Madison, USA, 1999.

Index

acoustic mixing, 7
  linear model, 7, 18, 149
  physical parameters, 8
adaptive interference canceler, 30, *see* AIC
  ILMS-adapted, 42
adaptive interference canceller, 163, 165
  ILMS-adapted, 165, 166, 172, 175
array steering, 31
assumption
  far- and free-field propagation, 15, 32
  gaussian p.d.f., 71
  independence, 40, 65
  instantaneous mixtures, 17, 74, 114, 116
  linear model, 149
  narrowband signal model, 108
  no target leakage, 45, 46
  sparse signals, 148
  square systems, 75
  stationarity, 12, 19, 39, 69
  steered microphone array, 31
  time-invariance, 10
  two sources, 78
  unit diagonal mixing channel, 21, 122
  white signal, 46, 47
  white signals, 123
  zero-mean signals, 10

background noise, 4, 55, 168, 176
  SNR, 168
beamforming, 3, 125, 142
  concept, 27
  prior knowledge, 27, 38

beampattern, 16
blind beamforming, 125
blind source separation, 3
  ambiguities, 3, 66
  blind, 63
  combined with beamforming, 147, 163
  compared to beamforming, 125
  convolutive mixtures, 120
  cost function, 113
  deflation approach, 77
  frequency domain, 108, 148, 175
  parameter settings, 101, 102
block-wise adaptation, 40, 97, 134, 143
blocking matrix, 165
  BSS-adapted, 166

car environment, 2, 149, 157, 175
causality of the mixing/separation
  system, 60, 95, 102, 111, 123, 151, 159, 163
  as prior information, 150
  delay-and-sum beamformer, 154
  permutation ambiguity, 96, 126
  source-microphone arrangement, 96, 175
cocktail-party problem, 1
compact microphone array, 2, 15, 23
complexity, 74, 78, 79, 91, 99, 133, 138
continuous adaptation, 39, 147, 165, 167, 172, 173
  implicit adaptation control, 42, 51, 61, 62

Printed in the United States
154176LV00004B/65/P